T0325699

Liquid–Gas and Solid–Gas Separators

There are no such things as applied sciences,
only applications of science.
Louis Pasteur (11 September 1871)

Dedicated to my wife, Anne, without whose unwavering support, none of this would have been possible.

Industrial Equipment for Chemical Engineering Set
coordinated by
Jean-Paul Duroudier

Liquid–Gas and Solid–Gas Separators

Jean-Paul Duroudier

First published 2016 in Great Britain and the United States by ISTE Press Ltd and Elsevier Ltd

ISTE Press Ltd
27-37 St George's Road
London SW19 4EU
UK

www.iste.co.uk

Elsevier Ltd
The Boulevard, Langford Lane
Kidlington, Oxford, OX5 1GB
UK

www.elsevier.com

For information on all our publications visit our website at http://store.elsevier.com/

British Library Cataloguing-in-Publication Data
A CIP record for this book is available from the British Library
Library of Congress Cataloging in Publication Data
A catalog record for this book is available from the Library of Congress
ISBN 978-1-78548-181-9

Printed and bound in the UK and US

Contents

Preface ix

Chapter 1. Separation by Decantation or Precipitation in a Magnetic or Electrical Field 1

1.1. Separation in a magnetic field 1
1.1.1. Recap on magnetic fields 1
1.1.2. Intensity of magnetization and magnetic separation 2
1.1.3. Classification of solids. 3
1.1.4. Processes of magnetic separation 4
1.2. Electrofilters 5
1.2.1. Description. 5
1.2.2. Shaking. 6
1.2.3. The transport mechanism 7
1.2.4. Apparition of sparks and counter-ionization 8
1.2.5. Geometric data 9
1.2.6. Operational equation 11
1.2.7. Calculation of the potential field 12
1.2.8. Consequences 15
1.2.9. Potential due to a uniform field and to a dielectrical particle. 15
1.2.10. Celerity of the ions 19
1.2.11. Charging of particles by ion bombardment 21
1.2.12. Charging of particles by heat diffusion 22
1.2.13. Smith and MacDonald's method 24
1.2.14. The classic expression of migration velocity 26
1.2.15. Shock re-entrainment of incident particles 28
1.2.16. Separation equations 30

1.2.17. By-pass effect 35
1.2.18. Variations in resistivity of the dust 36
1.2.19. How to modify the resistivity 41
1.2.20. Electrical power consumed and yield of a dust remover 42
1.2.21. Advantages of using an electrofilter 45
1.2.22. Usage 46

Chapter 2. Gas–Liquid Separator Vats and Drums 47

2.1. Gas separators and flash drums 47
2.1.1. Definitions 47
2.1.2. Gas separators 47
2.1.3. Entrainment rate 49
2.1.4. Horizontal flash drums (degassers) 52
2.2. Separator vats 54
2.2.1. Dimensioning 54
2.3. Conclusions 62
2.3.1. Liquid-vesicle particle sizes (estimation) 62
2.3.2. Vertical flash drums 64
2.3.3. Summary table 65

Chapter 3. Wet Dust Removal from Gases: Venturi Pulverization Column, Choice of a Dust Remover and Other Devices 67

3.1. The venturi 67
3.1.1. Description of a venturi 67
3.1.2. Usage of a venturi 69
3.1.3. Energy equation 69
3.1.4. Equation of state 71
3.1.5. Conservation of the flowrates 71
3.1.6. Impulsion of the gas and pressure drop in the convergent 72
3.1.7. Pressure drop due to the entrainment of the scrubbing liquid 73
3.1.8. Recovery of the pressure in the divergent 73
3.1.9. Size of drops of the scrubbing liquid 75
3.1.10. Separation equation 76
3.1.11. Individual yield of the drops 77
3.1.12. Motion of the drops 78
3.1.13. Length of the neck 79
3.1.14. Theoretical dust capture yield 79

3.1.15. Capture yield in practice 81
3.1.16. Integration limits for the integral I 81
3.2. Example of simulation of a venturi 82
3.2.1. Energy balance at the entrance to the convergent 82
3.2.2. Pressure at the entrance to the neck 84
3.2.3. Heat balance at the entrance to the neck 85
3.2.4. Pressure drop in the neck 86
3.2.5. Pressure at the outlet from the divergent 87
3.2.6. Diameter of the drops of scrubbing water 87
3.2.7. Length of the neck 87
3.2.8. Capture yield 88
3.3. Pulverization columns 89
3.3.1. Theoretical capture yield 89
3.3.2. Pressure drop of the gas in the column 93
3.4. Various points 96
3.4.1. Other types of wet scrubbers 96
3.4.2. General conclusion on wet scrubbing 97
3.5. Choice of an air scrubber 97
3.6. Varied calculations 98
3.6.1. Approximate calculation of the integral 98
3.6.2. Calculation of the overall heat balance for a wet scrubber 99

Chapter 4. Separation Between a Fluid and a Divided Solid Through Centrifugal Force 103

4.1. The cyclone 103
4.1.1. Principle and precautions for use 103
4.1.2. Phases and components 103
4.1.3. Presentation of the pressure drop 105
4.1.4. Pressure drop in the separation volume 105
4.1.5. Calculation of Ui/Vi (Barth's method, 1956) 106
4.1.6. Energy loss from the fluid upon return 109
4.1.7. Overall pressure drop 111
4.1.8. Pressure drop in hydrocyclones (approximate calculation) 113
4.1.9. Cunningham correction 115
4.1.10. Motion of dust particles 116
4.1.11. Residence time of fluid filaments 118
4.1.12. Capture yield for a given particle size 119
4.1.13. Simplified calculation of the yield of a gas cyclone 120

4.1.14. Simplified calculation of the yield of a hydrocyclone 122
4.1.15. Wet aerosols 124
4.1.16. Conclusions 124
4.2. The disk decanter 125
4.2.1. Description 125
4.2.2. Velocity profile 126
4.2.3. Theoretical separation yield 128
4.2.4. Evolution of the theoretical yield as a function of the flowrate 130
4.2.5. True separation yield 130
4.2.6. Engorgement 132
4.2.7. Use of disk centrifuge decanters 132
4.3. Tubular decanter 134
4.3.1. Description 134
4.3.2. The separation yield 134
4.4. Screw mud separator 137
4.4.1. Operational principles (Figure 4.9) 137
4.4.2. Velocity of the clarified liquid between the threads of the screw 139
4.4.3. Conclusions 141
4.4.4. Performances of the mud pump 142

Appendices 145

Appendix 1. Numerical Integration: the Fourth-order Runge–Kutta Method 147

Appendix 2. The Cgs Electromagnetic System 149

Appendix 3. Mohs Scale 153

Appendix 4. Definition and Aperture of Sieve Cloths 157

Bibliography 159

Index 163

Preface

The observation is often made that, in creating a chemical installation, the time spent on the recipient where the reaction takes place (the reactor) accounts for no more than 5% of the total time spent on the project. This series of books deals with the remaining 95% (with the exception of oil-fired furnaces).

It is conceivable that humans will never understand all the truths of the world. What is certain, though, is that we can and indeed must understand what we and other humans have done and created, and, in particular, the tools we have designed.

Even two thousand years ago, the saying existed: "faber fit fabricando", which, loosely translated, means: "*c'est en forgeant que l'on devient forgeron*" (a popular French adage: *one becomes a smith by smithing*), or, still more freely translated into English, "practice makes perfect". The "artisan" (faber) of the 21st Century is really the engineer who devises or describes models of thought. It is precisely that which this series of books investigates, the author having long combined industrial practice and reflection about world research.

Scientific and technical research in the 20th century was characterized by a veritable explosion of results. Undeniably, some of the techniques discussed herein date back a very long way (for instance, the mixture of water and ethanol has been being distilled for over a millennium). Today, though, computers are needed to simulate the operation of the atmospheric distillation column of an oil refinery. The laws used may be simple statistical correlations but, sometimes, simple reasoning is enough to account for a phenomenon.

Since our very beginnings on this planet, humans have had to deal with the four primordial "elements" as they were known in the ancient world: earth, water, air and fire (and a fifth: aether). Today, we speak of gases, liquids, minerals and vegetables, and finally energy.

The unit operation expressing the behavior of matter are described in thirteen volumes.

It would be pointless, as popular wisdom has it, to try to "reinvent the wheel" – i.e. go through prior results. Indeed, we well know that all human reflection is based on memory, and it has been said for centuries that every generation is standing on the shoulders of the previous one.

Therefore, exploiting numerous references taken from all over the world, this series of books describes the operation, the advantages, the drawbacks and, especially, the choices needing to be made for the various pieces of equipment used in tens of elementary operations in industry. It presents simple calculations but also sophisticated logics which will help businesses avoid lengthy and costly testing and trial-and-error.

Herein, readers will find the methods needed for the understanding the machinery, even if, sometimes, we must not shy away from complicated calculations. Fortunately, engineers are trained in computer science, and highly-accurate machines are available on the market, which enables the operator or designer to, themselves, build the programs they need. Indeed, we have to be careful in using commercial programs with obscure internal logic which are not necessarily well suited to the problem at hand.

The copies of all the publications used in this book were provided by the *Institut National d'Information Scientifique et Technique* at Vandœuvre-lès-Nancy.

The books published in France can be consulted at the *Bibliothèque Nationale de France*; those from elsewhere are available at the British Library in London.

In the in-chapter bibliographies, the name of the author is specified so as to give each researcher his/her due. By consulting these works, readers may gain more in-depth knowledge about each subject if he/she so desires. In a reflection of today's multilingual world, the references to which this series points are in German, French and English.

The problems of optimization of costs have not been touched upon. However, when armed with a good knowledge of the devices' operating parameters, there is no problem with using the method of steepest descent so as to minimize the sum of the investment and operating expenditure.

1

Separation by Decantation or Precipitation in a Magnetic or Electrical Field

1.1. Separation in a magnetic field

1.1.1. *Recap on magnetic fields*

In a vacuum, the magnetic induction caused by an electrical current I has the following dimensions:

$$[B] = \mu_0 [H] = \mu_0 \times \frac{\text{intensity}}{\text{m}}$$

$\vec{H}$: magnetic field

μ_0: magnetic permeability of a vacuum

The value of μ_0 depends on the system of units used.

International system (SI)	$\mu_0 = 4\pi.10^{-7}$	$[B]$ = Tesla or Ampere per meter $\times \mu_0$.
EMCGS	$\mu_0 = 4\pi$	$[B]$ = gauss

Let us examine the units of magnetic field:

1 oersted is the value of the magnetic field H which, in a vacuum, produces B = 1 gauss.

The SI unit of field, in a vacuum, produces an induction of $4\pi.10^{-7}$ Tesla.

By definition:

$$1\ \text{oersted} = \frac{1{,}000}{4\pi}\ \text{A.m}^{-1}$$

From this, we deduce:

$$1\ \text{A.m}^{-1} \rightarrow 4\pi.10^{-7}\ \text{Tesla}$$
$$1\ \text{oersted} \rightarrow 1\ \text{gauss}$$

We can divide both sides of the equation in turn, because the induction is proportional to the field.

$$\frac{1{,}000}{4\pi} = \frac{\text{oersted}}{\text{A.m}^{-1}} = \frac{\text{gauss}}{4\pi.10^{-7}\ \text{Tesla}}$$

Thus, we have the following correspondence between the units of magnetic induction:

$$1\ \text{Tesla} = 10{,}000\ \text{gauss}$$

1.1.2. *Intensity of magnetization and magnetic separation*

When subjected to a magnetic induction B, an elementary volume v of magnetizable substance acquires local magnetization M, where M is the magnetic moment of the elementary volume.

The volumetric intensity of magnetization per unit volume is:

$$J = \frac{\vec{M}}{v} = \chi \frac{\vec{B}}{\mu_0} = \chi \vec{H}$$

The dimensions of the magnetic moment are:

Electrical intensity $\times$surface = $I.m^2$

χ is the magnetic susceptibility, which is approximately 10^{-5} in the SI.

$$\text{As } [M] = I.m^2, \text{ we have } [M/v] = I.m^{-1}$$

J is therefore measured in the same units as H, and the magnetic susceptibility is a dimensionless number.

1.1.3. *Classification of solids*

1) Depending on the sign of the magnetic susceptibility:

– diamagnetics move away from zones of strong field, towards zones of weak field. Their susceptibility is negative (such is the case with graphite, for example). They cannot be separated using a magnetic field;

– ferromagnetics (Fe, Co, Ni), paramagnetics (Pt, Pd, Pyrite, FeO, MnO, NiO), antiferromagnetics and ferrimagnetics (ferrites M_2O_4 where M is a divalent metal such as Mg, Fe, Co or Mn) are attracted to zones where the field is strong. It is these substances that we can treat using a magnetic field.

2) Depending on their magnetizability:

– non-metallic minerals are not magnetizable, even by fields stronger than 25×10^5 $A.m^{-1}$;

– the separation of metallic minerals requires induction fields between 5×10^5 $A.m^{-1}$ (weak fields) and 25×10^5 $A.m^{-1}$ (strong fields). The range of susceptibilities is such that:

$$1{,}20.10^{-6} < \frac{\chi}{\mu_0} < 50.10^{-6} \qquad \text{(SI)}$$

(in the international system, the value χ/μ_0 is $1/4\pi$ CGS units of the value χ). Table 1.1 shows the susceptibility of a number of solids.

Note that magnetite, which is aptly named, can be separated by an induction field weaker than 1.2×10^5 $A.m^{-1}$ because, for it, $\chi/\mu_0 > 240\times10^{-6}$.

Finally, a solid is more magnetizable when its magnetic susceptibility is high.

Aegyrine	6.8	Hausmannite	4.6
Ankerite	2.7	Hornblende	2.4–18.3
Apatite	0.32	Ilmenite	2–9.1
Biotite	3.2–4.4	Manganite	3.8
Blende	0–0.8	Malachite	1.2
Bornite	0.6	Marcassite	0
Braunite	9.54	Mispickel	0
Calcite	0–0.24	Monazite	1.1
Manganiferous calcite	5.2–7.6	Psilomelane	2–4
Chalcocite	0	Pyrite	0
Corindon	0	Pyrolusite	–2.4
Dolomite	0.16	Quartz	0
Erubescite	2.15	Rhodocrosite	8–15.5
Feldspar	0.4	Rutile	42.16
Fluorine	0.4	Stilbine	0.48
Galene	0	Talc	2.23
Giobertite	0–1.2	Wad	6.8
Glauconite	4.8	Wernadite	2.4–3.58
Garnet	4.8–12.7	Wolframite	5.5–7.56
		Zircon	14

Table 1.1. *Susceptibility of minerals (χ/μ_0 in 10^{-6})*

1.1.4. *Processes of magnetic separation*

When a magnetizable solid is subjected to a magnetic induction field, the local magnetic moments are oriented in the direction of the induction fields, and the whole solid becomes similar to a magnet, each pole is attracted by the opposite pole of any magnet in the vicinity.

When we want to attract a product which is not highly magnetizable, we must use a powerful magnet, and *vice versa*. In the way of magnets, we could use:

– permanent magnets;

– electro-magnets, whose intensity can be varied from one machine to another;

– cylinders of mild steel, situated in the air gap of powerful magnets, and magnetized by induction ("induced rotors").

The general principle is to attract the magnetizable particles with a metal surface which is, itself, magnetized. Then, the metal surface moves out of the induction field and the particles that it had attracted detach, generally aided by centrifugal force (if the attracting surface is a cylinder) or a fluid current.

Whether the product is dry or dispersed in water, the principles remain the same. However, humid separation is limited by the viscosity of the liquid – particularly in the case of fine particles, whose velocity of displacement is reduced. Dry separation circumvents this problem but, in order to treat particles which are smaller than 20μm, the product's humidity must be less than 1%.

1.2. Electrofilters

1.2.1. *Description*

The general design of an electrofilter is shown by Figure 1.1. The wires have a lower potential than the plates. As the corona effect manifests around the wires, we say that, in this case, the discharge is negative.

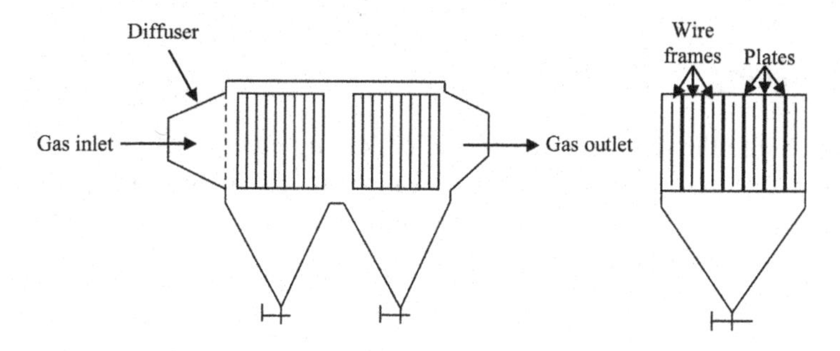

Figure 1.1. *Electrofilter*

The frames are held in place by quartz insulators (which are insensitive to significant variations in temperature), electrically heated if necessary, to prevent the formation of condensates on the insulators, because of the high water vapor content of the gases.

The shape of the receiver plates (collecting plates) is as shown in Figure 1.2.

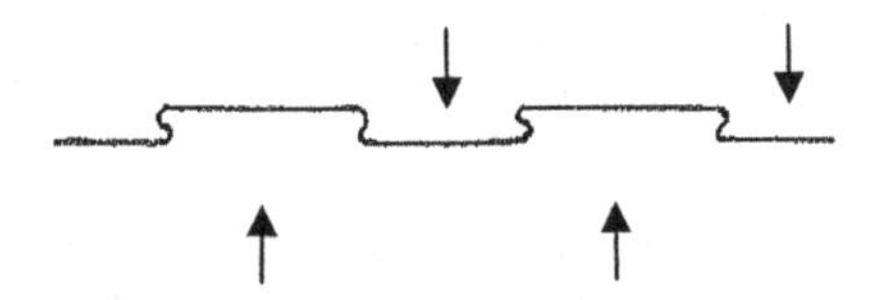

Figure 1.2. *Collecting electrodes*

This shape decreases the risks of the fixed dust particles rebounding. The velocity of the gases must be fairly high to render the installation profitable. However, it must not be too high, so as not to entrain the particles which have already precipitated.

The distribution of the gases as they enter the electrofilter is crucially important for the yield. The speed of the gas streams, measured with an anemometer, must be as uniform as possible. To this effect, we install perforated sheet metals in the inlet diffuser. These sheets create almost all of the pressure drop in the device. They constitute something of a pre-deduster, which necessitates the installation of a dust hopper beneath the diffuser.

When the gases enter through the top or the bottom, we install guide vanes rectifying the gaseous streams and bringing them to the horizontal in front of the perforated sheet metal at the inlet to the electrofilter.

1.2.2. *Shaking*

To dislodge the dust situated on the collecting electrodes, we periodically give those electrodes with a vibrational motion. The plates need to oscillate easily over the whole of their height so that the effect of the shaking at their bottom is transmitted fully to their top.

The shaking lends the plates an acceleration of 50 to 100 g ($g = 9.81\ m.s^{-2}$) and thus evacuates 90% of the dust collected.

The dust must fall in clusters because their drop speed is high (as the air resistance is negligible). Given the presence of the air stream, we cannot hope to make isolated particles fall. If such particles appear during the course of the shaking, the separation will need to be restarted from a distance of a few centimeters in front of the collecting electrode and, if the device is too short, the yield will drop.

Preferably, shaking should not be continuous and should be timed so that, between two operations, the clusters of dust have the time to grow – in other words, so that the thickness of the dust layer is sufficient. For example, we could adopt rest intervals of 5 minutes at the inlet and 15 minutes at the outlet (because it is here that there is least dust). The optimal interval between shaking cycles, and their intensity, need to be determined at startup.

If the dust has trouble detaching, we could spray:

– water if the dust is water-soluble;

– mineral oil for organic products (tar, naphthalene, etc.).

1.2.3. *The transport mechanism*

When there is a sufficient potential difference between a highly-curved electrode – say, a wire (and, *a fortiori*, a barbed wire) – and an electrode with little or no curvature like a plate, the wire is literally covered in charged ions which radiate in the ultra-violet spectrum. This is known as the corona effect.

In industrial electrofilters, the wires are given a negative potential in relation to the plates (the opposite is true in air-conditioning, so as to moderate the appearance of ozone). The wires constitute the emissive electrode and the plates (on which the dust accumulates) are the collecting electrodes.

Electrons are emitted by the wires by way of the photoelectrical effect, generally speaking, and in the electrofilter by the shock of positive ions moving away from the positive collecting electrode toward the wires. In their movement in opposite directions, the electrons encounter the molecules of gas (and the collisions between them liberate secondary electrons) and positive ions, with the production of UV radiation. Above all, though (and it is here that the interest lies), electrons affix to the gaseous molecules and transform them into negative ions, which in turn affix to the dust, giving it a negative charge.

1.2.4. *Apparition of sparks and counter-ionization*

The dust deposited on the collecting electrode has a certain resistance and, according to Ohm's law, a potential difference arises between the faces of the dust layer.

If the resistivity of the dust is sufficiently high, the corresponding potential difference may become greater than the breakdown voltage of the imperfect insulator, which is dust. Sparks fly (breakdown) and cause the formation of craters in which ions of both polarities come into being. This has a number of consequences:

1) when a spark appears, it is a preferential path for the electric current and the intensity passing through the installation increases sharply; it is then necessary for a device to be put in place to limit the intensity if we wish to avoid damaging the electrical circuits;

2) with the apparition of sparks, ions of both polarities are formed in the craters, and the positive ions formed are attracted into the gas space by the negatively-charged particles, equalizing and thus neutralizing their charge – hence the term "counter-ionization";

3) the formation of craters tends to re-disperse the dust collected into the gaseous environment.

The last two points above are significant causes of yield reduction and, in order for an installation to function properly, it is important that the frequency of disruptions is no higher than 1min^{-1} per electrical section. In practice, for the working voltage, we choose the nominal voltage compatible with that condition, which is tantamount to adopting a potential difference equal to 90% of the disruption voltage.

Note that the disruption electrical voltage is lower in the case of positive discharge than negative discharge, meaning that sparks occur more readily with a positive discharge. This is one of the reasons why dust removers operate with negative discharge.

1.2.5. *Geometric data*

Perforated sheet metal is installed at the input to the deduster so as to regulate the flow of gas. If these sheets are improperly attached, or if the fixings give way under the influence of a blow or of thermal expansion, an imbalance of gaseous flow will arise, which is detrimental to the overall yield, because the higher yield obtained for the slowest gaseous veins does not compensate for the degradation of the yield obtained with the faster veins.

If the wires were spaced a considerable distance apart, the current density on the collecting plates at the end of the wires could be 50% higher than the installation's average value. It is therefore useful to space the wires closer together – i.e. to increase the number thereof – but no great improvement is obtained once the distance separating two consecutive wires becomes smaller than the spacing between two plates.

The surface area of the collecting electrodes may be up to a maximum of 150 m^2 per $m^3.s^{-1}$ of gas in real conditions (non-standard). This surface area will be larger when separation is more difficult and, with the velocity of the gas remaining constant, the surface area determines the number of plates (i.e. their spacing) and the distance traveled by the gas in the device (characterized by the elongation – i.e. the ratio of the device's length to its height). Let us first examine this latter point.

At a given gas velocity V_G, the length of the gas path defines the residence time of the gas in the device, and that length of stay is no greater than 15 seconds. If the elongation is sufficient, natural sedimentation of the coarser particles occurs (particularly if V_G is no greater than 1 $m.s^{-1}$). The elongation also has an influence on the losses engendered by shaking (to dislodge the layer of dust from the plates). If the device is too short, dust falling from the plates may be entrained out of the electrofilter by the gas stream. In the case of finer dust, the drop time over a height of 7–10 m may reach up to several seconds and if, in addition, the elongation is less than 1 and V_G greater than 1.5 $m.s^{-1}$, there will be harmful consequences for the yield. In practice, the elongation L/H lies somewhere between the following bounds:

$$0.5 < L/H < 1.5$$

and, most usually:

$0.7 < L/H < 1.3$ (L: length; H: height)

The spacing of the plates varies, from case to case, between 0.15 and 0.40 m. Significant spacing is used when the dust concentration is high, so as to minimize the risks of sparks. On the other hand, spacing close together lowers the necessary voltage, but it is then more difficult to obtain a regular alignment. We shall soon see, though, that it is possible to circumvent this problem by dividing the device into independent sections.

A large surface of collecting electrodes with a single electrical inlet exhibits two weaknesses which favor the appearance of sparks because of imbalance of the current density:

– it is difficult to rigorously maintain the alignment of the electrodes;

– there are numerous local accumulations of dust not evacuated by the shaking.

Thus, the idea was to divide the device into sections, each with an independent electrical supply. Depending on the device, there may be between 2 and 8 sections per 100 $m^3.s^{-1}$ of gas, each section with a surface area ranging between 500 and 8000 m^2.

Thus, if we wish to process 100 $m^3.s^{-1}$ of a gas with 90 $m^2.(m^3.s^{-1})^{-1}$, we need to use 9000 m^2, which could be divided into three sections of 3000 m^2. By working in this way, we can hope to increase the voltage at which sparks begin to appear by 5–10 kV.

In summary, with a proper design, we can expect a good yield, because there will be:

– uniformity of the gas flowrate;

– correct alignment of the electrodes;

– uniformity of the current density on the electrodes.

If these three conditions are not respected, it is possible for a predicted yield of 95% to 99% to be, in reality, only 50% to 90%.

1.2.6. *Operational equation*

In the gaseous space, the flux of electrical charges passing across an elementary surface dS is:

$$\vec{n}\,(u\,q\,\vec{E})\,dS$$

Here, u is the mobility of the charged ions, E the electrical field and q the volumetric charge density. The vector n is the unit vector normal to the elementary surface dS.

In the permanent regime, there is no accumulation of charges in a closed surface:

$$\int_S \vec{n}(u\,q\,\vec{E})\,dS = 0$$

The mobility u is a constant which we can extract from the sum sign. Let us then apply Ostrogradsky's theorem:

$$0 = \int_S \vec{n}\cdot(q\vec{E})\,dS = \int_V \mathrm{div}(q\vec{E})\,dV$$

These equalities do not depend on the chosen volume; they are equivalent to:

$$\mathrm{div}\,(q\vec{E}) = 0$$

However, using vectorial analysis, we can show (see Spiegel, p. 120) that:

$$\mathrm{div}\,(q\vec{E}) = q\,\mathrm{div}\,\vec{E} + \vec{E}.\overrightarrow{\mathrm{grad}}\,q$$

According to Maxwell's laws:

$$q = \mathrm{div}\,\vec{D} = \varepsilon_r\,\varepsilon_0\,\mathrm{div}\,\vec{E}$$

(In the case of interest to us here, the relative permittivity ε_r is very close to 1, because we are dealing with gas at atmospheric pressure).

From the last three equations, it results that:

$$\frac{q^2}{\varepsilon_0} + \vec{E} \cdot \overrightarrow{\text{grad}}\, q = 0$$

However:

$$\vec{E} = - \overrightarrow{\text{grad}}\, V$$

where V is the electrical potential. We finally obtain the operational equation:

$$q^2 = \varepsilon_0\, \overrightarrow{\text{grad}}\, V . \overrightarrow{\text{grad}}\, q$$

1.2.7. *Calculation of the potential field*

For calculating the potential field, we have two equations at our disposal:

– Poisson's equation:

$$\Delta V + \frac{q}{\varepsilon_0} = 0$$

– the operational equation:

$$q^2 = \varepsilon_0\, \overrightarrow{\text{grad}}\, V . \overrightarrow{\text{grad}}\, q$$

In the y direction, the arrangement of the wires is periodic, with a spatial period d and the domain of length d exhibits two axes of symmetry. Therefore, we merely need to perform the calculations on the cross-hatched domain D.

This domain is divided into elementary squares. Figure 1.4 represents four of these squares whose side length is equal to a.

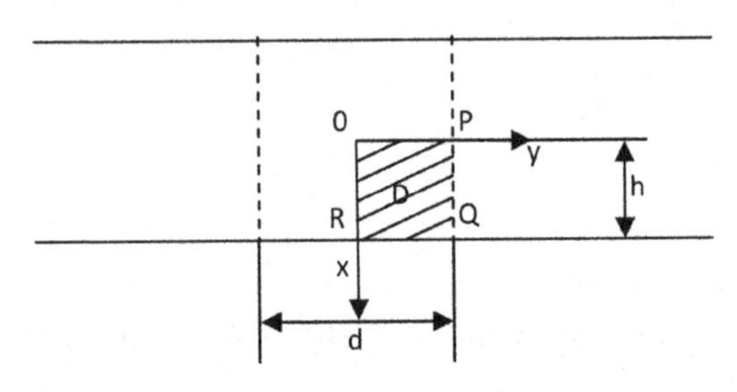

Figure 1.3. *Arrangement of wires and collecting electrodes*

For the potential, the boundary conditions are as follows:

Its value is imposed on the wire and on the collecting electrode. In addition, owing to the symmetry:

on the side OP: $\frac{\partial V}{\partial x} = 0$

on the sides OR and PQ : $\frac{\partial V}{\partial y} = 0$

Poisson's equation is written thus:

$$\frac{V_1 + V_2 + V_3 + V_4 - 4V_0}{a^2} + \frac{q_0}{\varepsilon_0} = 0$$

We then take a uniform charge density $q^{(0)}$ – e.g. equal to 10^{13} elementary charges per m^3 (i.e. 1.602×10^{-6} C/m^3).

The Poisson equation can then be used to calculate the potential field.

If we know the potential field, we need to calculate new values of the charge density. For this purpose, we need to know the lineic current density j_L on the wire.

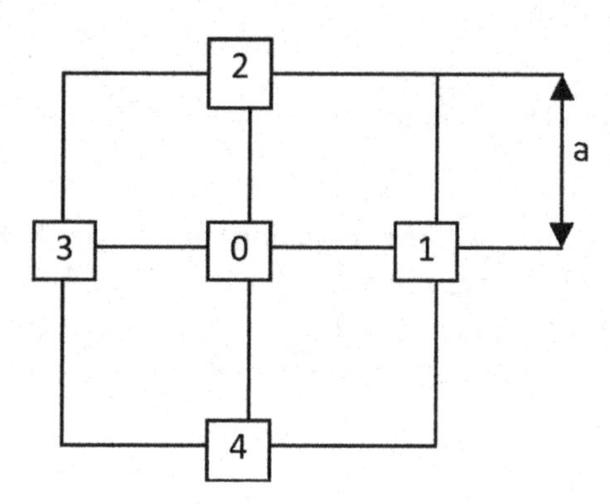

Figure 1.4. *Typical mesh*

If we suppose that the wire occupies the point 0 in the typical mesh, we can write the following, with u being the celerity of the charged particles:

$$\frac{j_L}{4} = a^2 u \left[\left[\frac{q_1 + q_0}{2} \right] \left[\frac{V_0 - V_1}{a} \right] + \left[\frac{q_4 + q_0}{2} \right] \left[\frac{V_0 - V_4}{a} \right] \right]$$

With the exception of the point 0, in light of symmetry, we are able to write:

on the side OP: $\frac{\partial q}{\partial x} = 0$

on the sides OR and PQ : $\frac{\partial q}{\partial y} = 0$

On the collecting electrode, we write the operational equation, expressing the derivatives with respect to x using the left-hand differences.

Within the domain D, we write that equation in the form:

$$q_0^{(1)} = \frac{\varepsilon_0}{4a^2 q_0^{(0)}} \left[(V_4 - V_2)(q_4 - q_2) + (V_3 - V_1)(q_3 - q_1) \right]$$

and on the collecting electrode, if we accept that the point 0 of the typical mesh is on that electrode:

$$q_0^{(1)} = \frac{\varepsilon_0}{q_0^{(0)}} \overrightarrow{\text{grad}}\, V \cdot \overrightarrow{\text{grad}}\, q$$

Thus, we have as many linear equations in $q^{(1)}$ as there are unknowns. After solving this system, we replace $q^{(0)}$ with $q^{(1)}$, and continue in this way until the values of q no longer vary. Having thus determined the range of the charge densities, we recalculate the potential range and continue until both ranges no longer vary.

Remember that the expression of the derivative by left difference is:

$$\frac{\partial V}{\partial x} = \frac{1}{2a} \left[3V_j - 4V_{j-1} + V_{j-2} \right]$$

1.2.8. *Consequences*

Leutert *et al.* [LEU 72] applied this method, and their publication gives the fields obtained. We then say that the electrical field is nearly uniform over four fifths of the distance separating the plates from the wires (starting at the plates). This is why, hereafter, we shall accept the hypothesis of a uniform electrical field at a distance from each particle which is significant in relation to the size of that particle.

On the other hand, as regards the electrical charges, the above calculations do indeed give the electrical charge at each point, but that charge is an overall charge which is the sum of three terms, pertaining respectively to:

– the electrons;

– the ions; or

– the charged particles of dust.

At present, we are unable to distribute the overall charge into those three elements. However, in the practical calculations, we accept that only the ions are significant.

1.2.9. *Potential due to a uniform field and to a dielectrical particle*

Let us begin by looking for the potential V_0 linked to the lone uniform field E_0. We can write:

$$-\frac{\partial V_0}{\partial x} = E_0 \qquad [1.1]$$

and

$$-\frac{\partial V_0}{\partial y} = 0 \qquad [1.2]$$

Let us take the origin of the potentials at point 0 and integrate equation [1.1]:

$$V_0 = -E_0 x$$

This expression automatically satisfies equation [1.2] and, finally:

$$V_0 = -E_0 r \cos\theta$$

Outside of the particle, which is initially supposed to be non-charged, we can seek an expression of the potential in the form of a serial expansion:

$$V_e = -E_0 r \cos\theta + \sum_{n=0}^{\infty} \frac{B_n}{r^{n+1}} (\cos\theta)^n$$

Indeed, a long way from the particle, we must find the expression of the potential due to a uniform field.

Inside of the supposedly-spherical particle whose center lies at 0, we can also look for an expression of the potential in the form of a serial expansion:

$$V_i = \sum_{n=1}^{\infty} A_n r^n (\cos\theta)^n$$

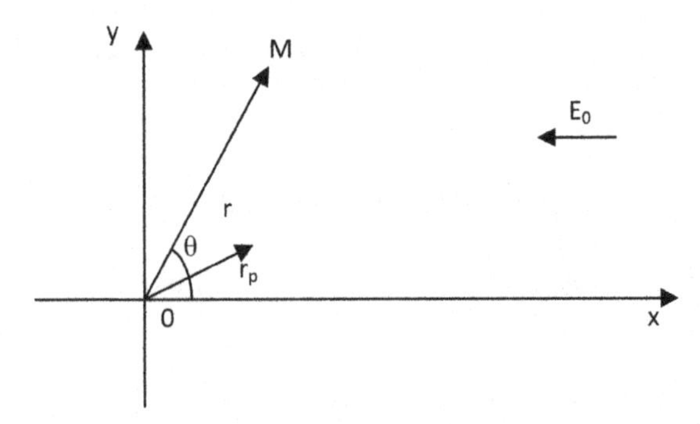

Figure 1.5. *Dielectric particle assimilated to point 0*

Indeed, at point 0, the potential must not only remain finite but actually disappear, because we have taken the origin of the potentials at that point.

Let us write that at the surface of the particle, the potential is continuous:

$$-E_0 r_p \cos\theta + \frac{B_0}{r_p} + \frac{B_1 \cos\theta}{r_p^2} + \frac{B_2 \cos^2\theta}{r_p^3} = A_1 r_p \cos\theta + A_2 r_p^2 \cos^2\theta \qquad [1.3]$$

Let us also write that the electrical induction is continuous:

$$\varepsilon_0 \left[\frac{\partial V_e}{\partial r}\right]_{r=r_p} = \varepsilon\,\varepsilon_0 \left[\frac{\partial V_i}{\partial r}\right]_{r=r_p}$$

$$-E_0\theta - \frac{B_0}{r_p^2} - \frac{2B_1\cos\theta}{r_p^3} - \frac{3B_2\cos^2\theta}{r_p^4} - \ldots = \varepsilon\left(A_1\cos\theta + 2A_2 r_p\cos^2\theta + \ldots\right) \qquad [1.4]$$

We can now identify the terms in cos θ in the above two equations.

Equation [1.3]	Equation [1.4]
$\frac{B_0}{r_p} = 0$	$\frac{-B_0}{r_p^2} = 0$
$-E_0 r_p + \frac{B_1}{r_p^2} = A_1 r_p$	$-E_0 - \frac{2B_1}{r_p^3} = \varepsilon A_1$
$\frac{B_2}{r_p^3} = A_2 r_p^2$	$\frac{3B_2}{r_p^4} = 2\varepsilon A_2$

By solving this set of equations, we obtain the following results:

$$B_0 = 0$$

$$A_1 = \frac{3E_0}{\varepsilon+2} \qquad\qquad B_1 = \frac{\varepsilon-1}{\varepsilon+2}E_0 r_p^3$$

$$A_2 = B_2 \cdots\cdots\cdots A_i = B_i \cdots\cdots\cdots = 0 \qquad (i \geq 2)$$

The external potential is therefore written:

$$V_e = -E_0 r\cos\theta\left[1 - \left[\frac{\varepsilon-1}{\varepsilon+2}\right]\frac{r_p^3}{r^3}\right]$$

To recap, the internal potential is written:

$$V_i = \frac{3E_0 r\cos\theta}{\varepsilon+2}$$

The angle θ is the angle of the field at infinity and of the local field.

It is easy to verify that these expressions satisfy Laplace's equation:

$$\Delta V = 0$$

If a charge q is uniformly distributed across the surface of the particle, the external potential becomes the sum of the potential we have just calculated and the potential due to that charge q:

$$V_e = \frac{q}{4\pi\varepsilon_0 r} + E_0 \cos\theta \left[r - \left[\frac{\varepsilon - 1}{\varepsilon + 2} \right] \frac{r_p^3}{r^2} \right]$$

Suppose that the positive ions approach the particle following the field lines which enter into that particle between the angles $\theta = 0$ and θ_ℓ close to it.

The limiting angle θ_ℓ is determined by the condition that the radial component of the field resulting from the above potential is zero. This is written:

$$-\left[\frac{\partial V}{\partial r} \right]_{r=r_p} = 0 = \frac{q}{4\pi\varepsilon_0 r_p^2} - E_0 \cos\theta_\ell \left[1 + 2\left[\frac{\varepsilon - 1}{\varepsilon + 2} \right] \right]$$

When $\theta_\ell = 0$, the ions can no longer approach the particle, and the corresponding value of the charge q is the value at saturation q_s:

$$q_s = 4\pi\varepsilon_0 E_0 r_p^2 \left[1 + 2\left[\frac{\varepsilon - 1}{\varepsilon + 2} \right] \right]$$

The charge q_s depends only on the strength of the field applied and the size of the particle.

When saturation is not reached, we have:

$$0 < q < q_s \qquad \text{but} \qquad \cos\theta_\ell = \frac{q}{q_s} \qquad \text{so } 0 < \cos\theta_\ell < 1$$

Consequently, the limiting angle of repulsion θ_ℓ is between 0 and $\pi/2$. For $\theta > \theta_\ell$, the particle tends to repel ions which approach it.

EXAMPLE 1.1.–

Consider ash whose relative permittivity is equal to 6, and suppose that the mean electrical field reigning between the wires and the collecting plate is 80,000 $V.m^{-1}$.

The charge of the particles at saturation is:

$$q_s = \frac{1}{9.10^9} \times 8.10^4 \times r_p^2 \left[1 + 2\left[\frac{6-1}{6+2}\right]\right]$$

$$q_s = 0.2\ 10^{-4} r_p^2 \qquad \text{(Coulombs)}$$

$$\frac{q_s}{e} = \frac{0.2 \times 10^{-4} r_p^2}{1.602 \times 10^{-19}} = 0.125 \times 10^{19} r_p^2$$

If we change r_p from 0.1 μm to 10 μm, we obtain the table below.

$r_p\,(m)$	10^{-7}	10^{-6}	10^{-5}
$q_s\,(C)$	$0,2.10^{-18}$	$0,2.10^{-16}$	$0,2.10^{-14}$
q_s/e	1.25	125	12500

Table 1.2. *Charge of particle*

1.2.10. *Celerity of the ions*

To calculate this celerity, we shall use the expression advanced by Langevin (1905). He used the CGS electrostatic system (see Appendix 2) in which celerity is expressed by:

$$v = \frac{3.10^{-2}}{16Y}\left[\frac{1}{(\varepsilon - 1)\rho_G}\right]^{1/2}\left[\frac{M+m}{m}\right]^{1/2}$$

ε: relative permittivity of the gas. For air: $\varepsilon = 1 + 6 \times 10^{-4}$

ρ_G: density of the gas ($g.cm^{-3}$)

M and m: molar masses of the gaseous molecules and of the ions. Practically:

$$M \# m \# 29 \text{ kg. kmol}^{-1} \text{ for air}$$

Langevin expressed the celerities in cm.s^{-1}. We shall express them, then, in m.s^{-1}.

To calculate 3/16Y, Langevin brings into play the factor:

$$\mu = \frac{e}{\sigma^2}\sqrt{\frac{\varepsilon - 1}{8\pi P}} \quad \text{(u.e.s. CGS)}$$

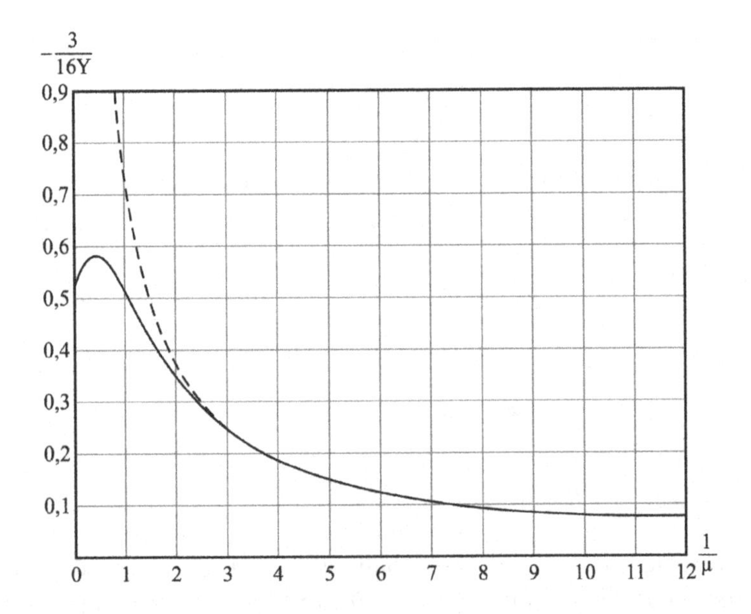

Figure 1.6. *Langevin curve for the mobility of a gaseous ion*

P: gas pressure (barye (1Pa = 10 baryes))

σ: collision diameter. For air: $\sigma = 0.362 \times 10^{-7}$ cm

e: charge on the electron (4.803×10^{-10} u.e.s. CGS)

For $\mu < 0.4$:
$$\frac{3}{16Y} = \frac{3\mu}{4}$$

For $\mu > 0.4$, we must read the value of 3/16Y on the curve in Figure 1.6 of Langevin [LAN 05]. This curve represents the variations of 3/16Y as a function of $1/\mu$.

Langevin's celerity v seems to be identical to the velocity of migration in electrofilters.

EXAMPLE 1.2.–

Let us examine the case of air in normal conditions:

$$P = 10^5 \text{ Pa} = 10^6 \text{ barye} \qquad \rho_G = 1.29.10^{-3} \text{ g.cm}^{-3}$$

$$\mu = \frac{4,803.10^{-10}}{0.362.10^{-7})^2}\sqrt{\frac{6.10^{-4}}{8\pi \times 10^6}}$$

$$\mu = 0.057$$

From the curve, we read the following for $1/\mu = 17$:

$$3/16Y = \frac{3 \times 0.057}{4} = 0.042$$

This, we have the celerity:

$$v = \frac{0.042}{\sqrt{6.10^{-4} \times 1.29.10^{-3}}} = 47.7 \text{ cm.s}^{-1}$$

$$v = 0.48 \text{ m.s}^{-1}$$

1.2.11. *Charging of particles by ion bombardment*

Following the field lines in the direction of decreasing potential, ions arrive at the particle and distribute themselves across its surface. Pauthenier and Moreau-Hanot [PAU 32] established the following expression for the rate of increase of the deposited charge:

$$\frac{dq}{d\tau} = \frac{q_s}{\tau_0}\left[1 - \frac{q}{q_s}\right]^2$$

N_0: spatial concentration of ions (m^{-3})

The other notations are the same as before.

The above equation can easily be integrated and, if we accept that q = 0 for $\tau = 0$, we obtain:

$$q = q_s \frac{\tau}{\tau + \tau_0} \quad \text{with} \quad \tau_0 = \frac{4 \times 2\,h}{v}$$

h: half-distance between electrodes of opposite signs: m

v: celerity: $m.s^{-1}$

EXAMPLE 1.3.–

$v = 0.48\ m.s^{-1}$ $\qquad h = 0.075\ m$

$$\tau_0 = \frac{4 \times 0{,}15}{0{,}48} = 1{,}25\ s$$

Thus, after 0.5 s:

$$\frac{q}{q_s} = \frac{0{,}5}{0{,}5 + 1{,}25} = 0{,}29$$

1.2.12. *Charging of particles by heat diffusion*

The electrical field is not the only factor to have an influence. Simply due to thermal agitation, it is possible for ions to come into contact with the particle. According to Maxwell–Boltzmann statistics, the rate of increase of the charge due to this phenomenon is linked to the interaction potential of the particle, whose radius is r_p and whose charge is q, with the ion whose charge is e. That potential is:

$$U = \frac{qe}{4\pi\varepsilon_0 r_p}$$

Thus, according to White [WHI 55]:

$$\frac{dq}{d\tau} = N_0 \pi r_p^2 \bar{V} \exp\left[\frac{-U}{kT}\right]$$

This equation is easy to integrate and, if we accept that q = 0 for τ = 0, we obtain:

$$q = \frac{4\pi\varepsilon_0 r_p kT}{e} \text{Ln}\left[1 + \frac{r_p \bar{V} N_0 e^2 \tau}{4\pi\varepsilon_0 kT}\right]$$

$\bar{V}$: effective velocity of thermal agitation of the ions ($m.s^{-1}$)

$$\bar{V} = \sqrt{\frac{3RT}{m}}$$

R: ideal gas constant (8314 $J.kmol.^{-1} K^{-1}$)

m: molar mass of the ions: $kg.kmol^{-1}$

T: absolute temperature (K)

k: Boltzmann's constant (1.38×10^{-23} J. $(molecule)^{-1}$)

The other notations are the same as before.

EXAMPLE 1.4.–

$$\tau = 0.5 \text{ s} \qquad r_p = 10^{-6} \text{m} \qquad T = 273 \text{ K}$$

$$m = 29 \qquad N_0 = 2.9.10^{13} \text{m}^{-3} \qquad 4\pi\varepsilon_0 = (9.10^9)^{-1} \text{C.m}^{-1}.\text{V}^{-1}$$

$$\bar{V} = \sqrt{\frac{3 \times 273 \times 8{,}314}{29}} = 485 \text{ m/s}$$

$$q = \frac{1.38.10^{-23} \times 273 \times 10^{-6}}{9.10^9 \times 1{,}602.10^{-19}} \text{Ln}\left[1 + \frac{10^{-6} \times 485 \times 2.9.10^{13} \times \left(1{,}602.10^{-19}\right)^2 \times 0.5}{1/\left(9.10^9\right) \times 1.38.10^{-23} \times 273}\right]$$

$$q = 2.61.10^{-18} \text{Ln}\left(1 + 431\right) = 1.58.10^{-17} \text{C}$$

We have seen that $q_s = 0.2\times10^{-16}$ (see section 1.2.9, Table 1.2); thus, we find:

$$q/q_s = 1.58.10^{-17}/0.2.10^{-16} = 0.79$$

Hence, we see that the roles of electrical field and thermal diffusion are of comparable importance, which led Smith and MacDonald [SMI 75] to propose a calculation method taking account of both phenomena simultaneously, rather than independently of one another, as we did above.

1.2.13. *Smith and MacDonald's method*

These authors divide the space into three regions depending on the value of θ:

Region I: $0 < \theta < \theta_\ell$

In this region, the kinetics resulting from the influence of the field is predominant and the charging takes place by ion bombardment:

$$\frac{dq_I}{d\tau} = \frac{N_0 e u q_s}{4\varepsilon_0}\left[1 - \frac{q}{q_s}\right]^2$$

Region III: $\frac{\pi}{2} < \theta < \pi$

In this region, the particle always, though weakly, repulses the ions, and only heat diffusion is active (with a coefficient of ½ because we are dealing with a hemisphere):

$$\frac{dq_{III}}{d\tau} = \frac{N_0 \pi r_p^2 \overline{V} e}{2}\exp\left[\frac{-qe}{4\pi\varepsilon_0 r_p kT}\right]$$

Region II: $\theta_\ell < \theta < \pi/2$

In this region, there is competition between diffusion and electrostatic repulsion. The authors accept that Maxwell–Boltzmann statistics applies on the basis of the expression of the diffusion kinetics:

$$\frac{dq_{II}}{d\tau}=\frac{N_0\pi r_p^2\overline{V}}{2}\int_{\theta_\ell}^{\pi/2}\exp\left[-\frac{U_{II}}{kT}\right]\sin\theta d\theta$$

The potential U_{II} is the sum of three terms:

– a pure diffusion term:

$$\frac{qe}{4\pi\varepsilon_0}\left[\frac{1}{r_p}-\frac{1}{r}\right]$$

– a term expressing the presence of the external field:

$$-V_{E_0}e$$

– a complementary term:

$$\frac{3r_pE_0\cos\theta e}{\varepsilon+2}$$

Finally:

$$\frac{dq}{d\tau}=\frac{dq_I}{d\tau}+\frac{dq_{II}}{d\tau}+\frac{dq_{III}}{d\tau}$$

This means that:

$$\frac{dq}{d\tau}=\frac{N_0euq_s}{4\varepsilon_0}\left[1-\frac{q}{q_s}\right]^2+\frac{N_0\pi r_p^2\overline{V}}{2}\exp\left[\frac{-qe}{4\pi\varepsilon_0r_pkT}\right]$$

$$+\frac{N_0\pi r_p^2\overline{V}}{2}\int_{\theta_\ell}^{\frac{\pi}{2}}\exp-\left[\frac{qe\left(r-r_p\right)}{4\pi\varepsilon_0kTr_pr}+\frac{\left[3r_pr-r^3\left(\varepsilon+2\right)+r_p^3\left(\varepsilon-1\right)eE_0\cos\theta\right]}{kTr\left(\varepsilon+2\right)}\right]\sin\theta d\theta$$

Let us now examine the numerical calculation.

At a given time, the particle carries a charge q, which allows the limiting angle of repulsion θ_ℓ and, when θ is between θ_ℓ and $\pi/2$ (zone II), the value of r involved in the integral is deduced by:

$$E = -\frac{\partial V}{\partial r} = 0 = \frac{q}{4\pi\varepsilon_0 r^2} - E_0 \cos\theta\left[1 + 2\left[\frac{\varepsilon - 1}{\varepsilon + 2}\right]\right]$$

Thus, here, r is the radial distance for the angle θ such that the radial component of the field is zero.

Finally, it is easy to numerically integrate the equation:

$$\frac{dq}{d\tau} = f,(q, \theta_\ell, r)$$

The charge reaches its saturation value q_s in around one second. This is the reason why the dust, if it is not re-entrained, is deposited near to the inlet of the electrofilter. Indeed, the length of stay in such a device varies from 3 to 12 seconds depending on the velocity chosen for the gas and on the length of the separators.

1.2.14. *The classic expression of migration velocity*

Using Stokes' law and applying the Cunningham correction to it, we obtain the theoretical expression of the migration rate:

$$w = \frac{qE}{6\pi\mu r_p}\left(1 + 0{,}86\frac{\lambda}{r_p}\right)$$

μ: viscosity of the gas (Pa.s)

r_p: radius of the particles (m)

λ: mean free path of the gaseous molecules (m)

λ is given by the expression:

$$\lambda = \frac{1}{\sqrt{2}} \cdot \frac{1}{\pi\delta^2} \cdot \frac{kT}{P}$$

δ: diameter of collision of gaseous molecules:

for air: $\delta = 0.362 \times 10^{-9}$ m

k: Boltzmann's constant (1.38×10^{-23} J.K^{-1})

T: absolute temperature (K)

P: pressure (Pa)

EXAMPLE 1.5.–

$T = 200°C = 473$ K $\quad P = 10^5$ Pa $\quad r_p = 10^{-5}$ m

$E = 2 \times 10^5$ V.m^{-1} $\quad \mu = 25 \times 10^{-6}$ Pa.s

$$\lambda = \frac{1}{\sqrt{2}} \times \frac{1}{\pi\left(0{,}362.10^{-9}\right)^2} \times \frac{1{,}38.10^{-23} \times 473}{10^5} = 0{,}19.10^{-7}\,\text{m}$$

Thus, we obtain the table below:

r_p(m)	10^{-7}	10^{-6}	10^{-5}
q_s(C)	0.5×10^{-18}	0.5×10^{-16}	0.5×10^{-14}
$6\pi\mu r_p$	0.47×10^{-10}	0.47×10^{-9}	0.47×10^{-8}
$1+0.86\dfrac{\lambda}{r_p}$	1.16	1.016	1.0016
w (m/s)	2.46×10^{-3}	2.16×10^{-2}	2.13×10^{-1}

Table 1.3. *Migration velocities*

1.2.15. *Shock re-entrainment of incident particles*

The flux density of particles coming onto the deposited dust layer is Cw.

Each incident particle upsets those which have already been deposited and caused the expulsion of some of them. More specifically, each incident particle expels e particles.

Let rw be the velocity of the particles ejected in a direction normal to the collecting electrode.

The kinetic energy of the incident particles must:

– endow the e ejected particles (whose mass is m_p) with an amount of kinetic energy equal to:

$$\frac{1}{2}m_p r^2 w^2$$

– overcome the force Eq_s exerted by the electrical field on a length equal to the diameter d_p of the particles, which corresponds to the energy:

$$Eq_s d_p$$

The energy consumption is then written:

$$\frac{1}{2}m_p w^2 = e\left[Eq_s d_p + \frac{1}{2}m_p r^2 w^2\right]$$

Thus:

$$e = \frac{e_0}{1 + r^2 e_0}$$

where:

$$e_0 = \frac{m_p w^2}{2Eq_s d_p}$$

However:

$$m_p = \rho_s \frac{\pi}{6} d_p^3 \text{ and } w = \frac{Eq_s C}{3\pi\mu d_p} \qquad \text{(see 1.2.14)}$$

Hence:

$$e_0 = \frac{\rho_s E q_s C^2}{108\pi\mu^2}$$

The corresponding flux density is then:

$$\phi_R = Cwer$$

The resulting flux density is:

$$Cw(1-er)$$

This is tantamount to assigning the theoretical migration rate of the coefficient (1–er).

EXAMPLE 1.6.–

$\rho_s = 2000 \text{ kg.m}^{-3}$	$E = 2\times10^5 \text{ V.m}^{-1}$	$r = 5$
$q_s = 0.5\times10^{-14}$ Coulomb	$C = 1$	$\mu = 25\times10^{-6}$ Pa.s

$$e_0 = \frac{2000\times2.10^5\times0.5.10^{-14}\times1}{108\times\pi\times\left(25.10^{-6}\right)^2} = 9.43$$

$$e = \frac{9.43}{1+5^2\times9.43} = 0.04$$

$$\frac{\varnothing_R}{Cw} = er = 0.04\times5 = 0.20$$

$$(1\text{-}er) = 0.8$$

NOTE.–

We have just seen that e is much less than 1. The result of this is that a particle can only be ejected by a cluster containing 1/e incident particles (25 particles in the present case). This phenomenon therefore only occurs where there is accumulation of the dust in the gas space.

1.2.16. *Separation equations*

1) The linear law (law I):

Let us assume that, as soon as it enters the device, all the dust approaches the collecting electrode at the migration velocity w.

The zone situated near to the wires is soon cleared of dust (this is zone I) and zone II (in the vicinity of the collecting electrodes) has a constant concentration equal to the initial concentration C_0.

As the gas progresses through the device, zone I expands whilst zone II shrinks.

As the velocity V_G of the gas and the migration velocity are constant, the boundary separating the two zones is a straight line whose inclination on the x axis is θ:

$$\text{tg}\,\theta = \frac{y}{x} = \frac{w}{V_G}$$

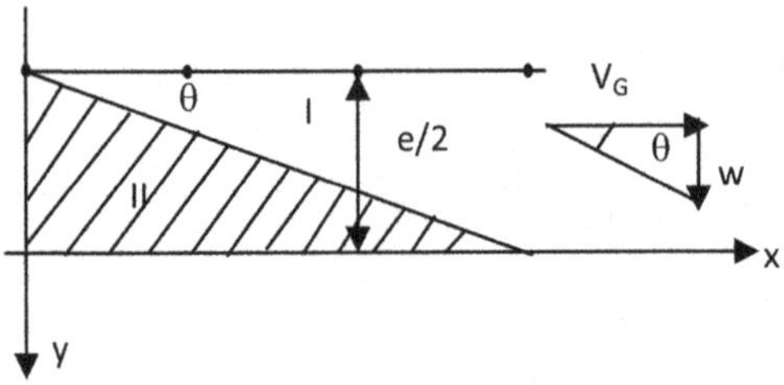

Figure 1.7. *Linear law*

At the abscissa point x, the mean concentration of the gas is:

$$\bar{C} = C_0\left(1 - \frac{y}{e/2}\right) = C_0\left(1 - \frac{xw}{(e/2)V_G}\right)$$

Let H represent the height of the collecting electrodes:

The collecting surface is: S = xH

The gas flowrate is: $Q_G = V_G\,(e/2)\,H$

Thus, we obtain:

$$\bar{C} = C_0\left[1 - \frac{Sw}{Q_G}\right]$$

This law does not accurately reflect reality, whilst Deutsch law, which we shall now examine, represents it more closely.

2) Deutsch law (law II):

We accept that the gas remains homogeneous in concentration regardless of the quantity of dust separated, and that concentration is:

$$\bar{C}\left(kg.m^{-3}\right)$$

The flux of dust deposited on the elementary surface dS is:

$$dS\bar{C}w$$

This flux corresponds, for the gas, to the loss:

$$-Q_G d\bar{C}$$

Hence:

$$\frac{d\bar{C}}{\bar{C}} = -\frac{wdS}{Q_G}$$

The boundary conditions are:

$$S = 0 \qquad \bar{C} = C_0$$

and, by integrating:

$$\bar{C} = C_0 \exp\left(-\frac{wS}{Q_G}\right)$$

The separation yield is:

$$\eta = 1 - \frac{\bar{C}}{C_0} = 1 - \exp\left(-\frac{wS}{Q_G}\right)$$

3) Taking account of re-entrainment (law III):

The flux of dust deposited becomes:

$$dS\left(\bar{C}w - \varnothing_E\right)$$

$\varnothing_E$ is the flux re-entrained (in kg.m^{-2}.s^{-1}).

The density is the same as for Deutsch's law and, after integration, we obtain:

$$\bar{C} = C_0 - C_0\left(1 - \frac{\varnothing_E}{wC_0}\right)\left[1 - \exp\left(-\frac{wS}{Q_G}\right)\right]$$

The yield becomes:

$$\eta = 1 - \frac{\bar{C}}{C_0} = \left(1 - \frac{\varnothing_E}{wC_0}\right)\left[1 - \exp\left(-\frac{wS}{Q_G}\right)\right]$$

4) Taking account of turbulent diffusion (law IV):

At a given point in the gaseous space, the flux density of dust in the direction of the collecting electrode is:

$$J = wC - D.\frac{\partial C}{\partial y}$$

D is the turbulent diffusion coefficient. In the absence of correlation for a flow in a channel with rectangular cross-section, we propose to use the relation pertaining to a circular cross-section but adapt it. This is Davies' relation:

$$D = \frac{0.01 V_G e}{Re}$$

e: space between the collecting plates (m)

V_G: velocity of the gas ($m.s^{-1}$)

$$Re = \frac{2W_G}{\mu H}$$

W_G: gaseous flowrate in terms of mass (kg/s (between two collecting plates))

H: height of the plates (m)

μ: viscosity of the gas (Pa.s)

The gas velocity being constant, the elapsed time is:

$$\tau = x/V_G$$

x is the distance traveled by the gas in the device and y is the vertical coordinate.

The continuity equation is tantamount to writing that the accumulation in an elementary volume is equal to the difference between the incoming and outgoing fluxes:

$$\frac{\partial C}{\partial \tau} = -\frac{\partial J}{\partial y}$$

so:

$$\frac{\partial C}{\partial \tau} = -w\frac{\partial C}{\partial y} + D\frac{\partial^2 C}{\partial y^2}$$

The boundary conditions are:

for $\tau = 0$ $C = C_0$ for any value of y

for $y = 0$ $\dfrac{\partial C}{\partial y} = 0$ (at the end of the wires, by symmetry)

for $y = \dfrac{e}{r}$ $\dfrac{\partial C}{\partial y} = 0$

The latter condition was put forward to prevent the appearance of a maximum concentration in the gaseous space.

This parabolic partial-differential equation can easily be integrated by a numerical method (see Nougier, p. 237). For this purpose, we merely need to know the migration rate w and the turbulent diffusion coefficient D.

Finally, Figure 1.8 shows the concentration profiles obtained for laws I, II and IV.

We can see that law IV is intermediary between laws I and II. Law III, which takes account of re-entrainment, holds no practical interest, because it would be unreasonable to run an electrofilter with sufficient gas velocity for there to be re-entrainment.

In conclusion, note that law IV, although it does obey logic, does not seem to have been studied by researchers.

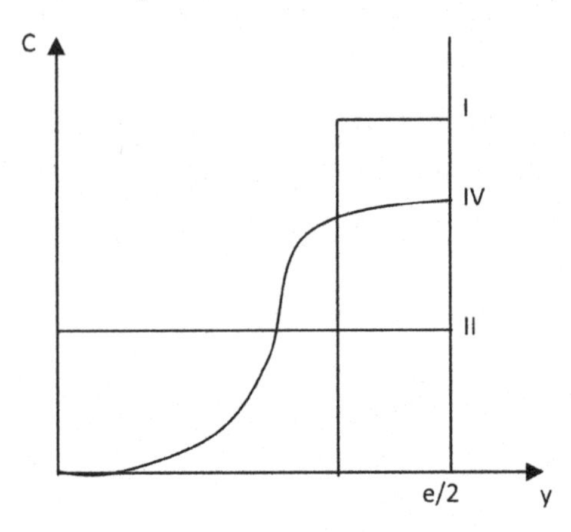

Figure 1.8. *Concentration profiles*

1.2.17. *By-pass effect*

The term “by-pass” has a very specific meaning. The prefix “bi” (anglicized as “by”) means that two passages are available to the gas:

– the normal passage;

– the parasitic passage through the hoppers (at the bottom) or through the insulation space for the frames supporting the wires (at the top).

This harmful effect can be remedied by the installation of chicanes. Between two chicanes, it is possible that 8-10% of the volume of gas is not treated within that section but, due to the effect of the chicanes, that volume is mixed back into the primary stream entering the next section.

Let B represent the volume fraction having circumvented the space of separation of section i. At the output from that section, the penetration is p_i, meaning that the unseparated fraction of dust is:

$$p_i = B + (1 - B)(1 - \eta_T)$$

η_T is the theoretical yield (without by-pass) of the section.

Now let us accept that the theoretical yields of all the sections are equal and have a common value η_T. For the whole of the device (which contains N sections), the real penetration will be:

$$p_{RE} = \left[B + (1 - B)(1 - \eta_T)\right]^N$$

If we have used a computer to calculate the yield η_{TE} of the whole device, supposed to have no by-pass effect, we have:

$$p_{TE} = 1 - \eta_{TE} = (1 - \eta_T)^N$$

or indeed:

$$1 - \eta_T = (1 - \eta_{TE})^{1/N}$$

Then, by eliminating η_T, we find that the true yield of the device is:

$$\eta_{RE} = 1 - p_{RE} = 1 - \left[B + (1-B)(1-\eta_{TE})^{1/N} \right]^N$$

EXAMPLE 1.7.–

$B = 0.08$ $N = 5$ $\eta_{TE} = 0.98$

$$\eta_{RE} = 1 - \left[0.08 + 0.92 \times 0.02^{1/5} \right]^5$$

$$\eta_{RE} = 0.968$$

1.2.18. *Variations in resistivity of the dust*

The resistivity of the dust bed is its dominant property, in terms of the electrostatic separation.

We measure the resistivity with the point device which enables us to simulate conditions similar to industrial conditions. A brief description of that device is to be found in White [WHI 74]. It includes an electrometer which must be stable and sensitive. It yields satisfactory results for resistivities greater than 10 Ω.m. Thus, in the following conditions:

surface of dust bed:	$0.001\ m^2$
thickness of dust bed:	0.0015 m
potential difference:	5000 V
current:	1.5×10^{-6} A

and therefore:

$$\rho = \frac{0.001}{0.0015} \times \frac{5000}{1.5.10^{-6}} = 2.2.10^9\ \Omega.m$$

Note that, in this case, the current density is equal to:

$$1.5.10^{-6}/0.001 = 1.5.10^{-3}\,\text{A.m}^{-2}$$

With a thermo-controlled casing, it is possible to operate at temperatures of up to 400°C.

The resistivity can be expressed in empirical terms as the resultant of two parallel conductances:

– a conductance at the surface of the particles: $1/\rho_s$

– a volume conductance within the grains: $1/\rho_v$

Hence:

$$\frac{1}{\rho} = \frac{1}{\rho_s} + \frac{1}{\rho_v} - \text{i.e. } \rho = \frac{\rho_s \rho_v}{\rho_s + \rho_v}$$

ρ_v increases with (1-ε), where ε is the porosity of the dust bed.

ρ_s increases with (1-ε)/d_p, where d_p is the mean particle diameter.

Thus, the stacking of the dust bed has an influence on the results of the measurements. The porosity ε lies between 0.7 and 0.8.

It is also possible to express the variation of ρ with temperature by writing:

$$\rho_s = A e^{-\frac{B}{T}} \text{ and } \rho_v = D e^{\frac{E}{kT}}$$

ρ_s decreases with the absorbed humidity and, if p_v is the partial pressure of water vapor in the gas, we can write:

$$\rho_s = A e^{\frac{-B' p_v}{T}}$$

It is easy to express the results of the measurements of ρ as a function of the temperature by an empirical expression with four parameters A, B', D and E. Figure 1.9 illustrates these variations.

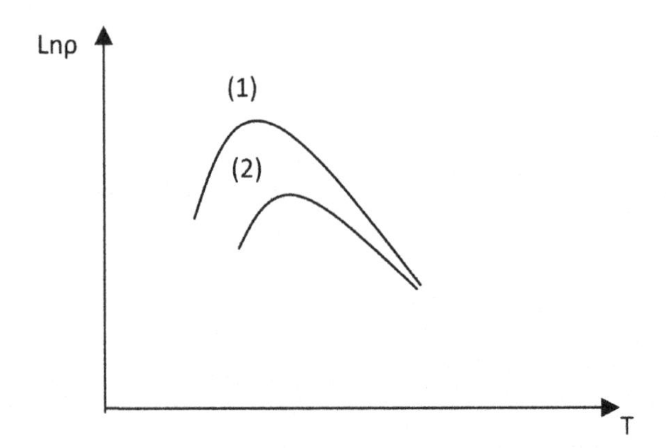

Figure 1.9. *Resistivity as a function of the temperature*

Curve (1) corresponds to a greater surface resistivity than curve (2). The position of the maximum depends heavily on the nature of the dust. It is given by Table 1.4.

In order for an electrofilter to work satisfactorily, the resistivity of the dust must lie between two limits:

– if the resistivity is too low, there will be re-entrainment;

– if the resistivity is too high, we see the formation of sparks.

Dust	ρ_{max}	T°C (approximate)
Anthracite (S: 2.5%; H_2O: 20%)	4×10^9	175
Quicklime	3×10^9	175
Cement	3×10^9	175
Chromium oxide	9×10^8	175
Alumina	2×10^{10}	120
Nickel oxide	3×10^8	20

Table 1.4. *Position of the maximum*

To express the effect of re-entrainment, we make reference to the work of Saludo *et al.* [SAL 80], and we shall use the semi-empirical theory of re-entrainment advanced in Duroudier, 1999.

These authors noted that for copper monoxide (CuO), re-entrainment occurred in the following conditions:

$$e = 0.16 \text{ m}; \rho = 3.10^6 \ \Omega.\text{m}; V_G = 1 \text{ m.s}^{-1}; d_p = 11 \ \mu\text{m}; j = 1.75.10^{-3} \text{ A.m}^{-2}$$

The flow is turbulent and corresponds to a pressure drop coefficient ? equal to 0.02 (see Brun *et al.*, Vol. II, p.70).

The velocity of entrainment is:

$$V_E = \frac{\rho_G V_G{}^2 d_p \psi}{16\mu} = \frac{1.3 \times 1^2 \times 1.1.10^{-5} \times 0.02}{16 \times 20.10^{-6}} = 0.0009 \text{ m.s}^{-1}$$

The electrical field is the product of the current density j by the resistivity ρ

$$E = 1.75.10^{-3} \times 3.10^6 = 5250 \text{ V.m}^{-1}$$

If, for particles of 11 μm, we take a charge of 10,000 elementary electronic charges, the force of cohesion will be:

$$F_C = Eq = 5.25.10^3 \times 10{,}000 \times 1.602.10^{-19} = 8.41.10^{-12} \text{ N}$$

The cohesion velocity is:

$$V_C = F_C / (3\pi\mu d_p) = \frac{8.41.10^{-12}}{3\pi \times 20.10^{-6} \times 1.1.10^{-5}} = 4.05.10^{-3} \text{m.s}^{-1}$$

The launching angle θ of the particles re-entrained is then given by:

$$\text{tg}\,\theta = V_E / V_C = \frac{0.0009}{0.00405} = 0.22$$

so:

$$\theta = 12.4 \text{ sexagesimal degrees}$$

However, there is a relation between the porosity of the dust bed and the angle θ:

$$\theta = 400\, e^{\frac{1}{\varepsilon - 1}}$$

That is: $\varepsilon = 0.71$ for $\theta = 12.4$ degrees

This value is perfectly acceptable for a dust bed.

Remember that the re-entrained flux density (in m^3/m^2s) is given by:

$$\varnothing_E = 4{,}5.10^{-4}\,(1-\varepsilon)\sin\theta\left(V_E\cos\theta - V_C\sin\theta\right)$$

The same authors verified that MgO was not re-entrained because its resistivity is 100 times greater. Consequently, E, F_C and V_C are also 100 times greater, and for the same porosity, the calculation yields a negative value for $\varnothing_E$ – in other words, its value is 0, because $\theta_E < 0$ makes no sense.

If the resistivity of the dust bed is too high, we see the appearance of sparks. The limiting value of the corresponding electrical field is:

$$1\times10^6 \text{ to } 2\times10^6 \text{ V.m}^{-1}$$

If we accept that the maximum value of the current density is always greater than 10^{-3}A.m^{-2}, we immediately obtain the upper limit of the resistivity:

$$\rho = \frac{E}{j} < \frac{E_{lim}}{j} = \frac{10^6}{10^{-3}} = 10^9\ \Omega.\text{m}$$

Manufacturers are cautious and prudent, and tend to accept 10^8 Ω.m as an upper limit of the resistivity.

1.2.19. *How to modify the resistivity*

Three scenarios need to be considered:

1) ρ_s is too low.

The overall resistivity is too low and there is re-entrainment. The surface resistivity is low if the humidity of the gas is too high (i.e. its dew point). Such a case may arise with the incineration of household waste, and the yield may drop from 0.98 to 0.90. We need to lower the dew point of the gas.

2) ρ_v is too low.

Here again, the overall resistivity is too low, and we see re-entrainment. Such is the case with coking charcoal particles by incomplete combustion in pulverized-charcoal heaters (graphite is a good electricity conductor). As the working temperature is higher than the temperature corresponding to the maximum of ρ, we need to lower that working temperature or, better still, work so that combustion is complete – i.e. increase the air excess.

3) ρ_s and ρ_v are both too high.

The value of ρ_{max} is often unacceptable, but as the working temperature of the device is beyond the maximum, the value of ρ is often satisfactory.

If this is not the case, we can:

– decrease ρ_v by increasing the temperature or incorporating 2–3% of caustic soda into the charcoal, as the sodium ions will take care of the electricity conduction;

– decrease ρ_s by increasing the humidity absorbed at the surface of the grains. To do so, we can incorporate 15 ppm of NH_3 into the gas, or 10 to 20 ppm of SO_3 if we are dealing with sulfur-free charcoal, or else, simply increase the humidity of the gas. Thus, ash from humid- and sulfurous charcoal has a sufficiently low resistivity.

1.2.20. *Electrical power consumed and yield of a dust remover*

1) Electrical characteristics.

The electrical energy needed to treat 1 m^3 of gas ranges from 100 to 1000 joules, and is usually between 200 and 400 joules. Compare this to the 4000 J needed for a humid washer to separate submicronic particles.

The intensity passing through the device is practically independent of the dust content of the gas, because the current is essentially transported by the negative ions.

The potential difference between the electrodes is between 30 and 100 kV, but the potential drop across the dust layer may be of the order of 1–5 kV, which decreases (by that amount) the voltage that can be used for ionization of the gas and transport of the dust. The usable voltage for the transport of the dust is:

$$\Delta V_u = \Delta V_T - j_S \rho h$$

ρ: resistivity of the dust layer (Ω.m)

j_S: current density on the collecting electrode ($A.m^{-2}$)

h: thickness of the dust layer (m)

EXAMPLE 1.8.–

Let us choose an electrofilter which is 7 m tall, 4 m wide and 7 m long.

With the collecting electrodes spaced 0.15 m apart, the surface area of the electrodes is:

Unitary surface × number of surfaces:

$$S = (7 \times 7) \times \left(2 \times \frac{4}{0{,}15} \right) = 2{,}613 \text{ m}^2$$

If the wires are spaced 0.15 m apart, the total length of wires installed is:

Height of wire × number of wires per channel × number of channels

$$L = 7 \times \left(\frac{7}{0.15}\right) \times \left(\frac{4}{0.15}\right) = 8{,}711 \text{ m}$$

If the mean migration velocity is 0.1 m.s^{-1}, the electrical power consumed by the device will be, as calculated using the simplified method:

$$P = \frac{2{,}613 \times 0.1}{0.04} = 6{,}532 \text{ W}$$

Suppose that the voltage applied between the electrodes is:

$$\Delta V_T = 35 \text{kV}$$

The intensity passing through the device will be:

$$I = \frac{P}{\Delta V_T} = \frac{6{,}532}{35{,}000} = 0.19 \text{ A}$$

The lineic current density on the wires will be:

$$j_L = \frac{I}{L} = \frac{0.19}{8{,}711} = 0.22.10^{-4} \text{ A.m}^{-1}$$

The current density on the collecting electrodes will be:

$$j_S = \frac{I}{S} = \frac{0.19}{2{,}613} = 73 \ \mu\text{A.m}^{-2}$$

Suppose that the dust layer is 1 cm thick and that its resistivity is 10^8 Ω.m. The usable voltage becomes:

$$\Delta V_u = 35{,}000 - 10^8 \times 73.10^{-6} \times 0.01 = 35{,}000 - 73$$

The influence of the dust is negligible. If the resistivity of the dust were 10^9 Ω.m, the corresponding potential drop would be around 1 kV.

Now let us accept that the mobility of the ions is:

$$u = 3.10^{-4} m.s^{-1} \left(V.m^{-1}\right)^{-1}$$

We can estimate the mean electrical field:

$$E = \frac{\Delta V_u}{(0.15/2)} = \frac{35,000}{0.075} = 467,000 \, V.m^{-1}$$

The mean concentration of charged ions, which are responsible for the electric current, will be:

$$N_0 = \frac{J_s}{euS} = \frac{73.10^{-6}}{1,602.10^{-19} \times 3.10^{-4} \times 467,000}$$

$$N_0 = 3.2.10^{12} \; m^{-3}$$

Suppose that, in order to satisfy the condition of non-re-entrainment, the velocity chosen for the gas is 1 $m.s^{-1}$, the gaseous flowrate passing through the device will be:

$$Q_G = 1 \times 7 \times 4 = 28 \; m^3.s^{-1}$$

We can then see that the specific surface of electrodes (per $m.^3s^{-1}$ of gas) is:

$$\frac{S}{Q_G} = \frac{2,613}{28} = 93 \; m^2.(m.^3s^{-1})^{-1}$$

Finally, the specific electrical power (per $m.^3s^{-1}$ of gas) is:

$$\frac{P}{Q_G} = \frac{6,532}{28} = 233 \; W.(m.^3s^{-1})^{-1} = 233 \; J.m^{-3}$$

2) Power consumed (simplified method):

There is a practical formula that can be used to evaluate that power:

$$P = \frac{Sw}{0.04} \qquad \text{(watt)}$$

S: surface of the collecting electrodes (m^2)

w: mean velocity of dust migration ($m.s^{-1}$)

In general, w is between 0.05 and 0.3 $m.s^{-1}$

If we know the potential difference ΔV between the wires and collecting electrodes, we can write:

$$P = \Delta V\, I$$

I is the intensity running through the wires.

When we know the total length L of the wires, we can write:

$$J_L = \frac{I}{L} = \frac{P}{L\Delta V}$$

J_L is the current intensity per meter of wire. This current density is measured in $A.m^{-1}$.

The intensity of the current passing through the deduster depends on the nature of the alloy of which the wires are made, the nature of the gas and also, of course, on the voltage applied.

3) Yield:

Gooch *et al.* [GOO 77] propose a mathematical model for the yield which uses the elements we have just discussed. This model consists of dividing the device into sections of 30 cm in length, and using the Deutsch equation in each section. The results obtained appear to closely match reality.

1.2.21. *Advantages of using an electrofilter*

The strengths of this device are as follows:

1) it can easily stop very fine particles $(0.1\,\mu m - 30\,\mu m)$. It is capable of processing many different aerosols. The temperature of the gas can reach 450°C with no problems;

2) its energy consumption is low: 150 to 800 joules per m^3 of gas processed. A humid washer to stop fine particles would consume ten times

more energy because, it would become highly energetic. In addition, the pressure drop on crossing the perforated plate situated at the inlet is negligible;

3) it is possible to decrease the gas flowrate to zero without the device's performances suffering.

1.2.22. *Usage*

Electrostatic dust removers are being used more and more regularly. Amongst their uses, we could cite:

– cement works;

– steel industry (agglomeration of iron ore);

– alumina manufacture and calcination kilns in general;

– coal- or household-waste-burning boilers;

– the operations of milling, drying and recovery of sulfuric- or phosphoric acid.

The gases or fumes issuing from these installations have a dust content of between 0.1 and 1 $g.m^{-3}$.

So-called "hot" devices are used especially to treat the gases issuing from calcination kilns and boiler fumes. In the latter case, they are situated before the air preheaters, which are therefore not fouled. The temperature of the gas treated in hot electrofilters is between 300 and 450°C. The resistivity of the dust is therefore low, which is favorable. However, the volumetric flowrate of gas is high, and its viscosity increases. Therefore, the risks of entrainment are no longer negligible, and the migration rate is lower.

2

Gas–Liquid Separator Vats and Drums

2.1. Gas separators and flash drums

2.1.1. *Definitions*

These are vessels designed to separate a gas and a liquid from a mixture obtained by expansion of a hot liquid. Expansion is achieved by causing the liquid to ascend in a vertical column (i.e. by reducing the hydrostatic pressure) or to escape through a valve.

Separator vessels fall into four categories:

– gas separators;

– vertical flash drums;

– horizontal flash drums (degassers);

– separator vats, which are large in diameter and generally have a conical bottom.

As we shall see later, it is important that the inlet to vats and degassers be immersed in the liquid.

2.1.2. *Gas separators*

These are simple cylindrically shaped, curved-bottom vessels, i.e. vertically mounted drums. The liquid-gas mixture enters the separator in the cylindrical part. The gas is discharged at the top, and the liquid at the bottom. Gas separators do not contain droplet-separating mats.

The diameter of a gas separator is usually determined by calculating the velocity of the gaseous phase V_{GV} in the empty, vertically mounted vessel:

$$V_{GV} = 0.06\sqrt{\frac{\rho_L}{\rho_G}} \qquad \left(m.s^{-1}\right)$$

ρ_L and ρ_G are the densities of the liquid and the gas expressed in $kg.m^{-3}$. The diameter D_S of the separator is derived from V_{GV}.

If the calculated velocity were the limiting velocity of a droplet in a highly turbulent flow regime, we could write:

$$g\rho_L \pi d_g^3/6 = 0.4 \times \frac{\pi d_g^2}{4} \times \frac{\rho_G V_{GV}^2}{2}$$

That is to say:

$$d_g = \frac{6 \times 0.4 \times 0.06^2}{8 \times 9.81} = 110 \times 10^{-6} m$$

In fact, the velocity V_{GV} corresponds to the intermediate flow regime, for which the drag coefficient is not equal to 0.4 but given by:

$$C_x = 18.5/Re^{0.6}$$

Hence, if we write that the weight is balanced by the drag force, we obtain:

$$196.4\ d_g^{1.6} = \left(\mu_G/\rho_G V_{GV}\right)^{0.6}$$

Gas separators only just capture raindrop-like droplets (around 500 μm in diameter). Nevertheless, they are very frequently used following the expansion of condensates of water vapor, since these are chemically pure species rather than mixtures, and the entrainment has no impact on the purity of the gas. Let D_S represent the diameter of the device.

This device satisfies the following dimensional constraints (which also apply to vertical separators in general):

1) the height of the liquid H_L must be such that its length of stay is approximately 3 min;

2) the height between the inlet and the level of the liquid must be at least equal to $0.5D_S$ – with an absolute minimum of 0.5 m – so that the incoming jet does not disturb the liquid surface and produce additional droplets that could be entrained;

3) the height separating the inlet from the bottom must be at least equal to D_S – with an absolute minimum of 1 m – so that the gas can change directions at a right angle and spread evenly over the whole area of the device;

4) if, at the inlet, we define:

$$\rho_e = \frac{W_G + W_L}{Q_G + Q_L} \quad \left(\text{kg.m}^{-3}\right) \quad \text{and} \quad V_e = \frac{Q_G + Q_L}{A_e} \quad \left(\text{m.s}^{-1}\right)$$

we should obtain:

$$\rho_e V_e^2 < 1500 \text{ kg.m}^{-1}.\text{s}^{-2}$$

This constraint prevents the creation of too many small vesicles, which would not be retained, and also prevents erosion of the wall.

Qs and Ws are volumetric and gravimetric flowrates, and A_e is the cross-section area of the inlet pipe;

5) it is recommended that the inlet be tangential rather than radial.

2.1.3. *Entrainment rate*

Let W_L (kg.s^{-1}) be the incoming liquid flowrate, and v the vaporization rate defined by:

$$v = \frac{W_G}{\left(W_L + W_G\right)}; \quad \text{thus,} \quad \frac{W_G}{W_L} = \frac{v}{1-v}$$

For practically all expansions, the vaporization ratio is such that:

$$0.01 < v < 0.15$$

Let C_E be the concentration of the liquid (in kg of liquid per m^3 of gas) entrained by the condensation discharged by the device:

for a gas separator: $C_E < 0.7 \times 10^{-3}$ kg.m^{-3}

with a droplet-separating mat: $C_E < 0.14 \times 10^{-3}$ kg.m^{-3}

When a mat is used, the entrained liquid flow is reduced by a factor of 5.

The flowrate of the gaseous phase, in terms of volume, is:

$$Q_G = W_G / \rho_G = \frac{W_L}{\rho_G}\left(\frac{v}{1-v}\right)$$

The entrainment rate is therefore:

$$E = \frac{W_E}{W_L} = \frac{C_E Q_G}{W_L} = \frac{C_E}{\rho_G}\left(\frac{v}{1-v}\right)$$

EXAMPLE 2.1.–

In a gas separator, we expand 2.8 kg.s^{-1} of water vapor condensate from 5 bar abs to 1 bar abs (100°C).

Using enthalpy tables, we can calculate the vaporization ratio v in terms of mass:

$$640.9 = (1-v)417 + v \times 2675$$

$$v = 0.0991$$

Thus, at 1 bar abs and at 100°C (saturation temperature):

condensation flow: 0.277kg.s^{-1} $\rho_G = 0.598$kg.m^{-3} i.e. 0.469 m^3.s^{-1}

solution flow: $2.523 kg.s^{-1}$ $\rho_L = 1000\ kg.m^{-3}$ i.e. $0.00262 m^3.s^{-1}$

Hence:

$$E = \frac{0.7 \times 10^{-3}}{0.598} \left(\frac{0,.0991}{1 - 0.0991} \right) = 0.013\%$$

$$V_{GV} = 0.06 \sqrt{\frac{1,000}{0.598}} = 2.45\ m.s^{-1}$$

Let us further adopt the hypothesis that:

$$\mu_G = 20 \times 10^{-6}\ Pa.s$$

$$d_g = \left[\left[\frac{20 \times 10^{-6}}{0.598 \times 2.45} \right]^{0.6} \frac{1}{196.4} \right]^{1/1.6}$$

$$d_g = 553 \times 10^{-6}\ m \quad (\text{raindrop})$$

The diameter of the separator would be:

$$D_S = \left[\frac{0.469}{2.45 \times (\pi / 4)} \right]^{1/2} = 0.494\ m \# 0.50\ m$$

The length of stay of the liquid would be 3 minutes:

$$0.00262 \times 3 \times 60 = H_L \times (0.469/2.45)$$

$$H_L = 2.46 m$$

Hence, the total height of the separator is:

$$H_E = 2.46 + 0.5 + 1 = 3.96\ m$$

$$\rho_e = 2.8/(0.469 + 0.00262) = 5.937\ kg.m^{-3}$$

The velocity after expansion though the valve should be such that:

$$5.937\ V_e^2 \leq 1500$$

That is:

$$V_e \leq 15.89\ \mathrm{m.s^{-1}}$$

Hence, we have the minimum diameter of the inlet pipe:

$$D_e \geq \left[\frac{0.4726}{15.89 \times (\pi/4)}\right]^{1/2} = 0.195\ \mathrm{m}$$

$$D_e \geq 0.2\ \mathrm{m}$$

2.1.4. *Horizontal flash drums (degassers)*

Like vertical drums, these devices are fitted with droplet-separating mats, but the free surface of the liquid is 4 to 8 times greater than in a vertical device of the same volume, thereby facilitating the degassing of the liquid – hence the name.

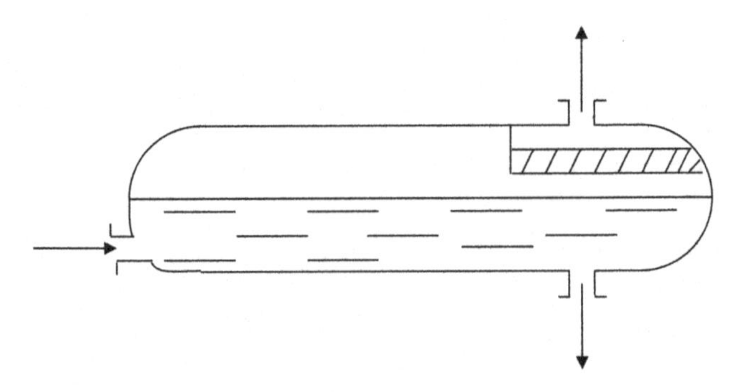

Figure 2.1. *Degasser*

The transverse area available for the flow of the liquid is the fraction R_L of the total section. The average value of R_L is 0.5. The surface area of the droplet-separating mat is $0.2\ D_S^2/4$, where D_S is the diameter of the separator.

The gas flow, in terms of volume, through the droplet-separating mats is:

$$Q_G = \frac{R_G \pi D_S^2}{4} K \sqrt{\frac{\rho_L}{\rho_G}}$$

If L_S is the length of the device, the ratio L_S/D_S is its elongation a, where the value of a is between 3 and 6.

As far as bubbles are concerned, the device behaves like a gravity decanter whose feed flow is:

$$Q_L = V_B L_S D_S = V_B D_S^2 a$$

With the *a priori* data of length of stay 5 min. and elongation a, D_S can be derived from Q_L.

V_B is the limiting ascension velocity of the bubbles separated by the device. It is given by Stokes' equation:

$$V_B = \frac{g\, D_B^2 \Delta\rho}{18\, \mu_L} \qquad \left(m.s^{-1}\right)$$

D_B: diameter of the bubbles (m)

$\Delta\rho$: difference in density between the liquid and the gas ($kg.m^{-3}$)

μ_L: viscosity of the liquid (Pa.s)

Hence:

$$D_S = \sqrt{\frac{Q_L}{V_B \times a}}$$

This device must satisfy the following dimensional constraints:

1) the liquid's length of stay must be such that:

$$\tau = \frac{\pi(1-R_G) D_S^2 L_S}{4 V_B D_S L_S} = \frac{\pi R_L D_S}{4 V_B} \geq 5\text{min., i.e. } R_L \geq R_L^* = \frac{4 V_B}{\pi D_S} \times 300$$

2) liquid filling rate:

$$R_L \# 0.5$$

3) elongation:

$$3 < a < 6$$

4) the inlet pipe must remain immersed at a depth of 0.1 m when the liquid is at its lowest level, to create a hydraulic guard.

5) the inflow must not be tangential to the ferrule. For further details, refer to mechanical construction codes.

EXAMPLE 2.2.–

$D_B = 50 \times 10^{-6}$ m $\rho_G = 1$ kg.m^{-3}

$Q_L = 14.7$ m^3.h^{-1} $= 0.00408$ m^3.s^{-1} $\rho_G = 1000$ kg.m^{-3}

$a = 3$ $K = 0.06$

$R_L \# 0.5$ $\mu_L = 0.001$ Pa.s

$$V_B = \frac{1000 \times 9.81 \times 25 \times 10^{-10}}{18 \times 10^{-3}} = 0.00136 \text{ m.s}^{-1}$$

Hence:

$$D_S = \left[\frac{0.00408}{3 \times 0.00136}\right]^{1/2} = 1 \text{ m and } R_L^* = \frac{4 \times 60 \times 5 \times 0.00136}{\pi \times 1} = 0.51$$

$R_L \# 0.5 \geq 0.51$ is verified in practice.

2.2. Separator vats

2.2.1. *Dimensioning*

Separator vats have a tangential inlet which is immersed in the liquid. They are designed to remove the liquid vesicles from the gaseous phase when the liquid is a solution of a *solid* product, as can be found in crystallization processes, and more generally, vaporization/concentration processes.

Indeed, solid deposits can form in the vapor pipes and, if the entrained vesicles remain liquid, they could cause erosion. Additionally, if the condensation is compressed, the compressor could be damaged. With an immersed inlet, the liquid vesicles entering the device are, for the most part, captured by the body of liquid present in the device, and the gas escapes in the form of bubbles. Separator vats rarely include a droplet-separating mats, since the mat would quickly become clogged by the solid deposits.

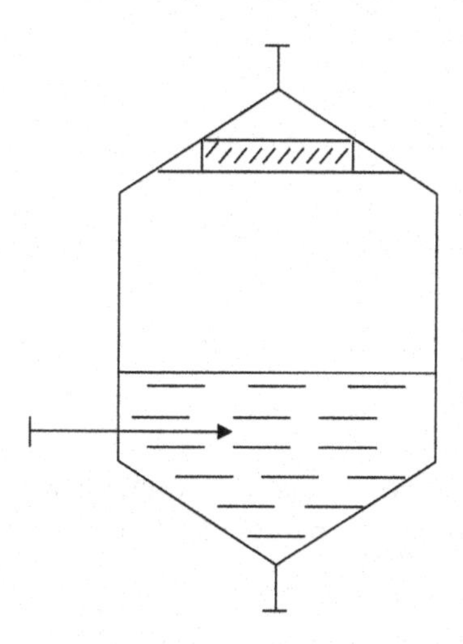

Figure 2.2. *Separator vat fitted (unusually) with a fiber droplet-separating mat*

As we have seen, the inflow of the liquid–gas mixture is tangential, and the gaseous fraction which escapes is impelled in a circular motion in the liquid. Its trajectory will therefore become helical, but the rotational movement will slow down and eventually cease when the gas escapes through the exhaust pipe at the top of the device.

Thus, two phenomena occur in the separation process:

– entrainment of the liquid vesicles by the gas, with ascension velocity V_{GV};

– centrifugal separation.

The velocity V_{GV} of the gas, which should be equal to the free-fall velocity of the vesicles, is obtained by using the equations corresponding to the intermediate regime (for which the Reynolds number lies between 0.3 and 1000).

Let us assume that the weight balances out the drag force:

$$\frac{\pi}{6}d_v^3\rho_L g = \frac{18.5}{Re^{0.6}} \times \frac{\pi}{4}d_v^2 \times \frac{\rho_G V_{GV}^2}{2}$$

where:

$$Re = \frac{V_{GV} d_v \rho_G}{\mu_G}$$

Hence, we have V_{GV} (and consequently the diameter D_S of the vat).

$$V_{GV} = 0.1528\left[\frac{d_v^{1.6} g \rho_L}{\mu_G^{0.6}\rho_G^{0.6}}\right]^{0.7143}$$

For D_S to permanently remain within reasonable limits when ρ_G decreases, V_{GV} is artificially increased by increasing d_V, which exceeds 200 µm at $\rho_G = 0.08$ kg.m^{-3}.

Let us now examine centrifugal separation. With this form of separation, the angular velocity which characterizes the helical movement of the gas, at the inlet, is:

$$\omega = V_{GH}/R_S$$

V_{GH} is the (horizontal) velocity of the gas in the inlet pipe, calculated as if it were alone, and R_S is the radius of the vat.

The angular velocity ω gradually diminishes and reaches zero when the gas escapes from the top of the separator vat. We will assume, therefore, that on average:

$$\bar{\omega} = \frac{1}{2}(V_{GH}/R_S) = V_{GH}/D_S$$

D_S is the diameter of the separator.

The radial velocity is given by the same equation as that used to calculate V_{GV}, provided that the acceleration due to gravity g is replaced with $\overline{\omega}^2 R$.

We therefore obtain:

$$\frac{dR}{d\tau} = A\,R^{0.7143}$$

where:

$$A = V_{GV}\left[\frac{\overline{\omega}^2}{g}\right]^{0.7143} = 0.1528\left[\frac{d_v^{1.6}\rho_L\overline{\omega}^2}{\mu_G^{0.6}\rho_G^{0.6}}\right]^{0.7143}$$

An integral gives the time τ required for a vesicle initially located at a radial distance of R_0 to travel the distance R_S i.e. to be pinned against the wall:

$$\tau = \frac{R_S^{0.2857}}{0.2857\ A}\left[1-\left(\frac{R_0}{R_S}\right)^{0.2857}\right]$$

Let us assume that the centrifugation was effective if:

$$R_S/R_0 = 10$$

This means that only a fraction of the gas equal to 1/100 (fraction of the horizontal section) will not be processed.

Hence:

$$\tau = 1.6872\frac{R_S^{0.2857}}{A}$$

or indeed:

$$\tau = 1.384\frac{D_S^{0.2857}}{A}$$

The height H_G of the gaseous space above the inlet is derived as follows:

$$H_G = \tau V_{GV}$$

More specifically, H_G is the vertical distance between the free surface area of the liquid and the top of the vat (where the vertical gas discharge pipe is located).

Furthermore, the design of a separator vat must satisfy certain design constraints:

1) to prevent excessive agitation of the immersed inlet from causing the formation of liquid vesicles that could be entrained, the horizontal velocity of the gaseous phase at the inlet, considered in isolation, must be such that:

$$\rho_G V_{GH}^2 < 20 \text{ kg.m}^{-1}.\text{s}^{-2}$$

2) it is recommended that a deflector be mounted inside the device to direct the jet towards the wall, by transforming the circular inlet into a rectangular inlet whose height is at least three times the width, and whose cross-section is equal to that of the circular inlet;

3) the diameter D_S of the vat should be equal to at least 6 times the diameter of the circular inlet;

4) the height H_G of the gaseous volume above the liquid surface should be greater than or equal to 2 m to enable the liquid filaments produced when the bubbles burst to trickle down;

5) the immersion height – i.e. the height of the liquid above the deflector – should be at least equal to 0.5 m;

6) the volume of liquid in the vat should correspond to a minimum length of stay of 5 minutes, so that the level can be controlled effectively;

7) there should be an anti-vortex (cross with vertical square branches) at the liquid outlet at the bottom of the vat to prevent gas entrainment if the level of liquid is at its minimum. The presence of gas in the liquid would be detrimental to the proper operation of an accelerating pump;

8) the height between the lower part of the deflector and the bottom of the vat should be at least 0.5 m.

EXAMPLE 2.3.–

Separation of vesicles from mist (water vapor) saturated at 50°C.

$W_G = 0.6\ kg.s^{-1}$ $\rho_G = 0.0831\ kg.m^{-3}$

$d_v = 350\times10^{-6}\ m$ $\mu_G = 20\times10^{-6}\ Pa.s$

$\rho_L = 1000\ kg.m^{-3}$ $Q_L = 0.0026\ m^3.s^{-1}$

At the inlet:

$$20 = 0.0831\ V_{GH}^2$$

$$V_{GH} = 15.52\ m.s^{-1}$$

$$A_e = \frac{0.6}{0.0831\times15.52} = 0.465\ m^2$$

Diameter of the inlet pipe:

$$D_e = \sqrt{\frac{0.465}{(\pi/4)}} = 0.77\ m \quad (DN\ 800)$$

$$V_{GV} = 0.1528\left[\frac{\left(350\times10^{-6}\right)^{1.6}\times9.81\times1000}{\left(20\times10^{-6}\right)\times0.0831^{0.6}}\right]^{0.7143}$$

$$V_{GV} = 3.65\ m.s^{-1}$$

Hence the diameter of the vat:

$$D_S = \left[\frac{0.6}{(\pi/4)\times0.0831\times3.65}\right]^{1/2} = 2.512\ m$$

$$\overline{\omega} = 15.52/2.512 = 6.178\ rad.s^{-1}$$

$$A = 3.65\left[\frac{6.178^2}{9.81}\right]^{0.7143} = 9.6327$$

$$\tau = 1.384\frac{2.512^{0.2857}}{9.6327} = 0.187\ \text{s}$$

$$H_G = 0.187 \times 3.65 = 0.68\ \text{m}$$

We shall therefore assume that:

$$H_G = 2\ \text{m}$$

$$H_L \times \frac{\pi}{4} \times (2.512)^2 = 0.0026(5 \times 60)$$

$$H_L = 0.157\ \text{m} \# 0.16\ \text{m}$$

The height of the deflector will be such that:

$$H_D = \sqrt{3 \times \frac{\pi}{4} \times 0.77^2} = 1.18\ \text{m}$$

Total height:

$$H_T = 2 + 0.68 + 0.5 + 1.18 + 0.5$$

$$H_T = 4.86\ \text{m}$$

The vertical velocity of the liquid is:

$$0.0026/\left(\frac{\pi}{4} \times 2.512^2\right) = 0.52 \times 10^{-3}\,\text{s}^{-1}$$

To evaluate the size limit of the separated bubbles, we shall express the ascension velocity of the bubbles using Stokes' law:

$$0.52 \times 10^{-3} = \frac{9.81 \times 1000 \times D_B^2}{18 \times 10^{-3}}$$

(We had previously assumed that the viscosity of the liquid was equal to that of water at 20°C, i.e. 10^{-3} Pa.s).

Hence:

$$D_B = 30 \times 10^{-6} \text{ m}$$

The bottom of the vat contains an "emulsion" of small bubbles whose diameter is most frequently between 1 and 5 mm; bubbles with the minimum diameter of 30 μm are discharged.

Let us ascertain that we are indeed operating in the laminar regime:

$$\text{Re} = \frac{30 \times 10^{-6} \times 5 \times 10^{-3} \times 1000}{10^{-3}} = 0.015$$

We can see that:

$$0.015 < 0.3$$

REMARK.–

As a general rule, separator vats are larger than gas separators and flash drums. Indeed, the latter devices will only separate vesicles with a diameter in excess of ~500 μm. Another reason is that separator vats are generally used in vacuum processes where ρ_G is low. If there are any 200 μm particles remaining in the condensation mists, no erosion will occur either in the pipe or, above all, in the exchanger or evaporator tubes, where these mists would be the heating fluid, provided that the velocity at which the condensation mists come into contact with the bundle is reduced by a divergent or a distribution ring.

On the other hand, if we want to re-compress the vapor using a centrifugal compressor, or a dry screw compressor (i.e. whose screws are not lubricated with oil so as not to pollute the mists), the vesicles should be no larger than 10 μm, and the vat should contain a fiber droplet separator.

Obviously, the vesicles should not clog the mat (especially if they contain dissolved chemical species).

2.3. Conclusions

2.3.1. *Liquid-vesicle particle sizes (estimation)*

To express these particle sizes, we shall use a modified version of the Rosin–Rammler law. Indeed, this law has the advantage of being analytically simple. Moreover, it describes the distribution of airborne dust in calcination and drying kilns. In our case, this law will describe the particle size of the entire body of incoming liquid but, of course, only the fraction containing liquid vesicles less than 1mm in size will be of interest in terms of liquid–gas separation. Indeed, the limiting free-fall velocity of a droplet with a diameter of 1mm is already far greater than the admissible gas velocities for both liquid–gas separating devices and fractionating columns (distillation, absorption, stripping).

The original form of the Rosin–Rammler equation is:

$$P(d_v) = \exp\left[-\left[\frac{d_v}{d_v^*}\right]^n\right]$$

$P(d_v)$ is the "pass-through" in the context of particle-size analysis using the sieve method. It is simply the gravimetric fraction of the vesicles whose diameter is less than d_v.

In a separating device, the size of the liquid vesicles cannot exceed the incoming diameter D_e of the feed tube. This size limit corresponds to the arrival of a liquid plug. Consequently, the modified law that we should consider is:

$$P^*(d_v) = A\,P(d_v)$$

In this equation, d_v is expressed in µm and we ascertain that:

$$P^*(d_v) = 1 \quad \text{and} \quad d_v = D_e$$

Indeed, the calculation shows that the coefficient A is very close to 1, and that this correction is mainly of interest from the point of view of theoretical consistency. The original Rosin–Rammler equation is therefore sufficient.

We shall now determine the parameters d_v^* and n for a free-flow inlet and for an immersed inlet. The magnitudes of dv and d_v^* are expressed in μm.

1) Free-flow:

Our hypotheses will be:

– a droplet-separating mat ceases to be effective at 10 μm, and the pass-through (entrained vesicles) is 0.1%;

– without a droplet-separating mat, a gas separator ceases to be effective at 500 μm, and the pass-through is 10%.

The resulting equation is:

$$P(d_v) = 1 - \exp\left[-\left(\frac{d_v}{3311}\right)^{1.19}\right] \quad [2.1]$$

In this equation, we can observe that an entrainment of 8% (which is often admitted for distillation columns) corresponds to a vesicle diameter of about 300 μm. This corresponds to a limiting free-fall velocity of a water droplet in air that is highly comparable to the vapor velocities found in perforated-plate distillation columns operating in the jet regime. Equation [2.1] is a reasonable estimation of the particle size in the entire body of liquid contained in the column (but for $d_v < 1$ mm).

2) Immersed inlet:

We will assume that the exponent 1.19 remains unchanged, but that d_v^* is now equal to 26,336 μm instead of 3,311 μm. This is the case in separator vats, and this value of d_v^* derives from the hypothesis that the entrainment is 0.3% for a vesicle size of 200 μm.

EXAMPLE 2.4.–

With the data from the previous vat-calculation example, the flowrate of the entrained liquid is:

$$0.3\% \times 2.60\ \text{kg.s}^{-1} = 0.0078\ \text{kg.s}^{-1}$$

The volumetric flowrate of the gas was:

$$0.6/0.0831 = 7.22 \text{ m}^3\text{s}^{-1}$$

Consequently, we obtain the following vesicle concentration:

$$0.0078/7.22 = 0.00108 \text{ kg.m}^{-3} = 1.08 \text{ g.m}^{-3}$$

This low concentration explains why it is generally not necessary for separator vats to have a droplet-separating mat.

NOTE.–

The above explains why vats are used in vacuum processes whereas flash drums – which handle concentrations that are often much higher – are used in processes where the pressure is greater than or equal to atmospheric pressure.

NOTE.–

The preceding "universal" law is simply an initial approach since, in reality, the particle size of the liquid vesicles depends on the densities ρ_L and ρ_G and, above all, on the velocity conditions at the inlet of the device, as well as on the surface tension of the liquid and the relative proportions of gas and liquid.

2.3.2. *Vertical flash drums*

Like gas separators, these drums have a free-flowing inlet – i.e. located above the surface of the separated liquid. They are used to separate a vapor and a liquid which are, themselves, mixtures of two or more fluids. These devices differ from gas separators in that they contain, in the top part of the device, a droplet-separating mat whose surface area is calculated using rules specific to this type of equipment, and the resulting diameter is adopted for the flash drum. The mat captures all particles whose diameter is greater than 10 µm.

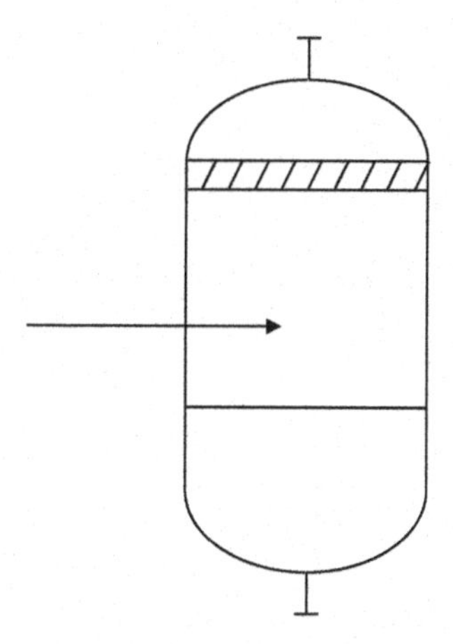

Figure 2.3. *Vertical flash drum*

Vertical flash drums must satisfy the same design constraints which apply to gas separators.

2.3.3. *Summary table*

In the degasser, the separated bubbles had a diameter of 30 μm. If we consider that the same holds true for a gas separator or a separator vat, the liquid flowrate in these devices must be 3 to 5 times less, as the free surface area of the liquid is 3 to 5 times smaller than in a degasser.

In the final analysis, given the examples presented, we can draw up the following summary table where d_v is the limiting diameter of the separated vesicles:

Type of device	d_V (μm)
Gas separator	500
Separator vat	200–350
Vertical flash drum	10
Degasser	10
Woven-fiber mat	10
Non-woven fiber mat	0.1

Table 2.1. *Limiting diameters*

3

Wet Dust Removal from Gases: Venturi Pulverization Column, Choice of a Dust Remover and Other Devices

3.1. The venturi

3.1.1. *Description of a venturi*

The principle of the device is to accelerate the gas by passing it through the neck of a convergent/divergent vessel. In the neck, this gas comes into contact with the initially immobile liquid, which resolves into drops which retain the dust brought in by the gas.

If the gas is hot – i.e. in practical terms, if its temperature is greater than the liquid's boiling point (100°C for water, which is the most commonly-used liquid) – it is necessary to cool it ("quench" it, to use the terminology of the field). To do so, we spray cold liquid at the entrance to the convergent. This cooling means we can conserve the material of the device and choose a material whose cost is moderate. In addition, the gas reaches saturation with vapor.

In the neck, the liquid is injected, for example, using a toric tube shot through with holes, because this liquid is often a recirculated suspension, containing suspended dust which would block a sprayer. The dust is then stopped by way of the inertia shock mechanism, meaning that under the influence of its inertia, the dust follows a broader trajectory than that of the fluid threads and thus comes into contact with the drops, whereas the gas avoids them.

In the divergent, compression is at work, and the vapor condenses onto the dust if that dust is at least partially soluble in the liquid. Thus, the dust particles serve as nuclei for the condensation of the vapor. In the liquor obtained by partial dissolution of the solid, the activity of the water is less than 1, and therefore the condensation takes place preferentially on the dust. Thus the dust particles become heavier and larger, making them easier to capture.

The device is all the more effective when the relative velocity of the gas with respect to the drops is high, which explains why venturis generally see the entrainment of the liquid by the gas, rather than the other way around. Indeed, the velocity of the gas at the neck is between 100 and 500 $m.s^{-1}$, whereas a liquid can only be injected at a velocity of the order of 10–25 $m.s^{-1}$.

The effectiveness of the venturi remains constant and independent of the gaseous flowrate if the gas encounters the liquid at a velocity independent of that flowrate. To make this possible, the inlet section has two lateral shutters whose inclination can be regulated to maintain a constant value for the pressure drop on passing through the device. Indeed, the depression at the neck is poorly recovered in the divergent section.

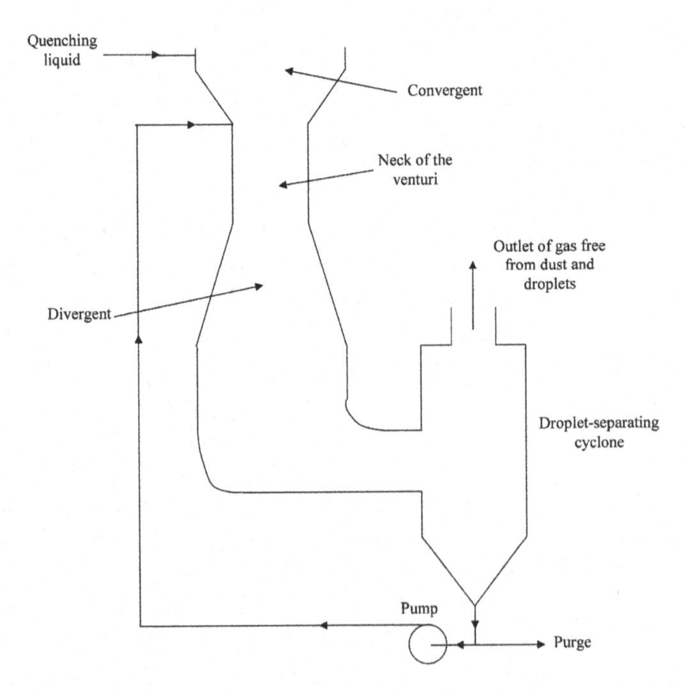

Figure 3.1. *Principle behind the venturi*

3.1.2. *Usage of a venturi*

The venturi is effective in stopping fine dust particles (<1 μm), provided the velocity of the gas in the device is sufficient, meaning that the corresponding pressure drop is sufficiently great.

This device is used to scrub combustion fumes and gases from certain ore-treatment plants (phosphate ore, for example) or fertilizer factories.

3.1.3. *Energy equation*

For the suspension of drops of liquid in the gas, we merely need to write that the sum of the enthalpies and kinetic energies of the gas and the liquid is a constant:

$$W_G H_G(T) + W_L h_L(T) + W_G E_{CG} + W_L E_{CL} = \text{const.}$$

W_G and W_L: mass flowrates of the gas and liquid ($kg.s^{-1}$)

H_G and h_L: gravimetric enthalpies of the gas and the liquid ($J.kg^{-1}$)

E_{CG} and E_{CL}: kinetic energies of the gas and the solid ($J.kg^{-1}$)

V_g is the velocity of the liquid or, more specifically, the velocity of the drops of liquid. u_G is the velocity of the gas.

At the inlet to the convergent, the liquid is pulverized in the direction of the gaseous flow at a variable velocity between 10 and 20 $m.s^{-1}$, and that velocity is oriented in the direction of progression of the gas. This moderate velocity is that of the gas in the inlet pipe. Therefore, the kinetic terms can be neglected in relation to the enthalpy terms, and thus we describe the total heat entering into the device:

$$q_E = W_{LE} h_L(T_{LE}) + W_{GE} H_{GE}(T_E)$$

However, the incoming gaseous flowrate W_{GE} is the sum of the flowrate W_I of the inerts and of the flowrate W_{VE} of the vapor of the same chemical nature as the liquid:

$$W_{GE} = W_I + W_{VE}$$

Thus:

$$q_E = W_I H_I (T_{GE}) + W_{VE} H_V (T_{GE}) + W_{LE} h_L (T_{LE})$$

With this equation, we can calculate q_E.

The thermal equilibrium is established almost instantaneously and at the mixing temperature T_M:

$$q_E = W_I H_I (T_M) + (W_{VE} + E) H_V (T_M) + (W_{LE} - E) h_L (T_M)$$

The flowrate E is exchanged between the gas and the liquid:

– if $E > 0$ there has been vaporization of liquid;

– if $E < 0$ there has been condensation of vapor.

By bringing in the mean heat capacities of the substances present, and the latent heat of vaporization of the liquid (L_0 at 0°C), the conservation of the energy is written:

$$q_E = E(T_M C_V - T_M C_L + L_0) + (W_I C_I + W_{V0} C_V + W_{L0} C_L) T_M + L_0 W_{V0}$$

That is to say:

$$E = \frac{q_E - T_M (W_I C_I + W_{V0} C_V + W_{L0} C_L) - L_0 W_{V0}}{T_M C_V - T_M C_L + L_0}$$

The index 0 characterizes the vapor- and liquid flowrates W_{V0} and W_{L0} after injection of the quenching liquid but still at the inlet to the convergent.

The mixing temperature T_M is the boiling point of the liquid at the pressure P_V:

$$P_V = P_E y_{V0} = P_E \frac{W_{V0} + E}{W_{V0} + E + W_I \dfrac{M_V}{M_I}}$$

P_E is the total pressure at the point where mixing takes place. M_V and M_I are the molar masses of the vapor and of the inerts.

Thus, the calculation takes place by successive iterations, choosing as the initial value of T_M the dew point of the gas arriving in the venturi. Another method will be found in Appendix II, where we express the gaseous flowrates in $Nm^3.s^{-1}$.

3.1.4. *Equation of state*

The equation of state can be written:

$$\rho_G = \frac{M_G P}{RT}$$

M_G: mean molar mass of the gaseous mixture ($kg.kmol^{-1}$)

P: total pressure of the mixture (Pa)

T: absolute temperature of the mixture (K)

R: ideal gas constant (8314 $J.kmol^{-1}.K^{-1}$)

For example, at the inlet to the convergent:

$$M_G = y_V M_V + (1 - y_V) M_I$$

$$P = P_E$$

Hence, we obtain ρ_G, which is the density of the gas ($kg.m^{-3}$).

3.1.5. *Conservation of the flowrates*

The liquid flowrate, in terms of mass, is written:

$$W_L = \rho_L V_g A_L$$

The gaseous flowrate, also in terms of mass, is written:

$$W_G = \rho_G u_G A_G$$

ρ_L and ρ_G: densities of the liquid and the gas (kg.m^{-1})

A_L and A_G: parts of the cross-section area occupied by the liquid and the gas (m^2)

V_g: velocity of the liquid drops (m.s^{-1})

u_G: velocity of the gas (m.s^{-1})

Only the second equation will be used in studying the motion, assimilating A_G to the total area. Indeed, the value of ρ_L is much higher than ρ_G, and the result of this is that A_L is a very small fraction of the total cross-section.

3.1.6. *Impulsion of the gas and pressure drop in the convergent*

Let us write that the loss in pressure of the gas creates the motion of the drops and accelerates the gas:

$$-AdP = W_G du_G + W_L dV_g \qquad (V_g \text{ is the velocity of the drops})$$

We shall accept the hypothesis that, in the convergent:

$$u_G = V_g = V = \frac{Q_G}{A} = \frac{W_G}{\rho_G A} = \frac{RTW_G}{M_G PA}$$

Thus:

$$dV = -\frac{RTW_G}{M_G PA^2} dA - \frac{RTW_G dP}{M_G AP^2}$$

and:

$$-PdP = -\left(W_{G0} + W_{L0}\right) W_G \frac{RT}{M_G}\left[\frac{dA}{A^3} + \frac{1}{A^2}\frac{dP}{P}\right]$$

$$P_E^2 - P_C^2 = \left(W_{G0} + W_{L0}\right) W_G \frac{RT}{M_G}\left[\frac{1}{A_C^2} - \frac{1}{A_E^2} + \frac{2}{\bar{A}^2} Ln\frac{P_E}{P_C}\right]$$

A_C and A_E: section of the neck and of the entrance to the convergent (m^2)

W_{G0} and W_{L0}: mass flowrates of the gas and liquid ($kg.s^{-1}$)

P_E and P_C: pressure at the entrance to the convergent and the entrance to the neck (Pa)

We have implicitly accepted a low temperature variation, remaining equal to T_M.

This equation in P_C is solved in two iterations, accepting that:

$$\frac{1}{\overline{A}^2} = \frac{1}{3}\left[\frac{2}{A_C^2} + \frac{1}{A_E^2}\right]$$

(At the first iteration, the logarithm is taken as equal to zero).

3.1.7. *Pressure drop due to the entrainment of the scrubbing liquid*

We have just seen that at the outlet from the neck, the liquid shifts practically at the velocity of the gas and, as shown in the description of the pneumatic transport of a divided solid in the dilute phase, the corresponding pressure drop of the gas is:

$$\Delta P_{EL} = \frac{W_{LD} u_G}{A_C}$$

W_{LD}: mass flowrate of scrubbing liquid ($kg.s^{-1}$)

u_G: velocity of the gas ($m.s^{-1}$)

A_C: cross-section of the neck (m^2)

3.1.8. *Recovery of the pressure in the divergent*

If the gas is alone, the recovery of pressure in the divergent varies from 50 to 70%. It is all the better as the divergent is longer, meaning that the angle which its walls form with the axis is small. On the other hand, the transformation of the kinetic energy of the solid into gas pressure takes place with a mediocre yield of the order of 10 to 15%. The majority of the energy is dissipated into friction.

If ΔP_G is the ideal recovery of the gas pressure considered as being alone, in order to find the overall recovery we must multiply ΔP_G by the yield η_0, which we shall define by:

$$\eta_0 = \frac{0.6W_G + 0.1W_L}{W_G}$$

To express the ideal recovery of the lone gas, we use the expression established for the convergent:

$$\Delta\left(P^2\right) = W_G^2 \frac{RT}{M_G}\left[\frac{1}{A_C^2} - \frac{1}{A_D^2}\right]$$

A_D: cross-section of outlet from the divergent (m^2)

If we now set, for the output pressure P_S from the divergent:

$$P_S = P_{SC} + \Delta P_D$$

P_{SC}: outlet pressure from the neck (Pa)

We obtain:

$$\Delta\left(P^2\right) = P_S^2 - P_{SC}^2 = \left(\Delta P_D\right)^2 + 2P_{SC}\Delta P_D \# 2P_{SC}\Delta P_D$$

The corrective multiplier coefficient η_0 will therefore be the same for ΔP_G and $\Delta(P^2)$. Hence:

$$\Delta\left(P^2\right) = \left(0.6W_G + 0.1W_L\right) W_G \frac{RT}{M_G}\left[\frac{1}{A_C^2} - \frac{1}{A_D^2}\right]$$

and:

$$P_S = \left[P_{SC}^2 + \Delta\left(P^2\right)\right]^{1/2}$$

3.1.9. *Size of drops of the scrubbing liquid*

Nukijama and Tanasawa [NUK 39] established the expression of the harmonic mean diameter of the drops resulting from the meeting of an almost-immobile liquid and a gas animated with a certain velocity in relation to the liquid:

$$d_g = \frac{0.585}{u_G}\left[\frac{\sigma}{\rho_L}\right]^{0.5} + 1.683\left[\frac{\mu_L^2}{\sigma\rho_L}\right]^{0.225}\left[\frac{Q_L}{Q_G}\right]^{1.5}$$

d_g: diameter of the drops (m)

σ: surface tension of the liquid ($N.m^{-1}$)

ρ_L: density of the liquid ($kg.m^{-3}$)

μ_L: viscosity of the liquid (Pa.s)

Q_L and Q_G: volumetric flowrates of the liquid and the gas ($m^3.s^{-1}$)

Water is generally the liquid used for the operation of the venturi. Its properties are:

$$\sigma = 0.072\ N.m^{-1} \qquad \rho_L = 1000\ kg.m^{-3} \qquad \mu_L = 10^{-3}\ Pa.s$$

The above expression is then written:

$$d_g = \frac{0.585}{u_G}\left[\frac{0.072}{1000}\right]^{0.5} + 1.683\left[\frac{10^{-6}}{0.072 \times 1000}\right]^{0.225}\left[\frac{Q_L}{Q_G}\right]^{1.5}$$

That is to say:

$$d_g = \frac{0.004964}{u_G} + 0.0288\left[\frac{Q_L}{Q_G}\right]^{1.5}$$

u_G is the velocity of the gas in an empty bed.

3.1.10. *Separation equation*

In the slice whose thickness is $d\ell$, with constant cross-section area A and volume $A d\ell$, the incoming flowrate of dust is:

$$cQ_G$$

The portion of this flowrate (sensible flowrate), in which the dust is likely to be stopped, must correspond to the frontal area of the liquid drops.

$$cu_G dA_F = cu_G a_F A dl$$

a_F is the total frontal area of the drops present in the unitary volume and c is the concentration of dust (kg.m^{-3}).

Let us introduce the volume of liquid β_L present in the unitary volume of dispersion:

$$\beta_L = \frac{A_L}{A} = \frac{1}{A}\left[\frac{Q_L}{V_g}\right] \quad \text{(dimensionless)}$$

Let us also introduce the value α_F, which is the ratio of the frontal surface $\pi d_g^2/4$ of a drop to its volume $\pi d_g^3/6$.

$$\alpha_F = \frac{3}{2d_g} \qquad \left(m^{-1}\right)$$

The frontal section of the drops contained in the volume $A d\ell$ is:

$$dA_F = \alpha_F \beta_L A dl = \frac{3Q_L}{2d_g V_g} dl$$

The probability that the dust will encounter drops in the volume $A d\ell$ is the ratio of the frontal cross-section of the drops to the total section available to the dusty gas:

$$\frac{dA_F}{A} = \frac{3Q_L \eta_u}{2d_g V_g A} dl$$

The gas threads separate on the passage of a drop, but because of inertia, the dust particles adopt a tighter trajectory and find the drop, which corresponds, for that drop, to the individual yield η_I, which is the ratio of the circular section surrounding the captured particles to the frontal cross-section of the drop. Finally, the relative variation of the concentration of the gas in terms of dust in the volume $A d\ell$ is:

$$-\frac{dc}{c} = \eta_I \frac{dA_F}{A} = \frac{3\eta_u Q_L}{2d_g V_g A} dl$$

We shall call this equation the separation equation.

3.1.11. *Individual yield of the drops*

We use Barth's approximate expression:

$$\eta_I = \frac{V_R^2}{V_R^2 + V_0^2}$$

with:

$$V_0 = \frac{1.144 \times 18 d_g \mu_G}{d_p^2 \Delta\rho} = \frac{20.6\ d_g \mu_G}{\Delta\rho d_p^2} \quad (m.s^{-1})$$

μ_G: viscosity of the gas (Pa.s)

d_g: harmonic mean diameter of the drops (m)

d_p: diameter of dust particles (m)

$\Delta\rho$: difference between the true density of the dust and that of the gas ($kg.m^{-3}$)

V_R: relative velocity of the gas in relation to the drops ($m.s^{-1}$)

$$V_R = u_G - V_g$$

3.1.12. *Motion of the drops*

The law of point dynamics is written:

$$m\frac{dV_g}{d\tau}=\frac{\pi}{6}d_g^3\rho_L\frac{dV_g}{d\tau}=C_x\frac{\pi d_g^2}{4}\frac{\rho_G V_R^2}{2}\qquad \left(dV_g>0\right)$$

with:

$$V_R=u_G-V_g$$

This means:

$$d\tau=\frac{4}{3C_x}\frac{\rho_L}{\rho_G}d_g\frac{dV_g}{V_R^2}$$

However, this motion takes place in the intermediary regime between the laminar and turbulent regimes. Thus, the drag coefficient C_x is written:

$$C_x=\frac{18.5}{Re^{0.6}}=18.5\left[\frac{\mu_G}{d_g\rho_G}\right]^{0.6}\frac{1}{V_R^{0.6}}$$

Hence:

$$d\tau=k_g\frac{dV_g}{V_R^{1.4}}=-k_g\frac{dV_R}{V_R^{1.4}}$$

with:

$$k_g=\frac{4\rho_L d_g}{3\times 18.5\times\rho_G}\left[\frac{d_g\rho_G}{\mu_G}\right]^{0.6}$$

The equation of motion is written as follows, taking account of the fact that $d\tau=d\ell/V_g$:

$$\frac{1}{k_g}=\int_{V_{g1}}^{V_g}\frac{V_g dV_g}{\left(u_G-V_g\right)^{1.4}}$$

The index 1 characterizes the inlet to the neck after injection of the scrubbing liquid. By integrating part-wise, we obtain:

$$l = k_g \left[\frac{u_G - 0.4V_g}{0.4 \times 0.6 \left(u_G - V_g\right)^{0.4}} \right]_{V_{g1}}^{V_g}$$

3.1.13. *Length of the neck*

We admit that, when the velocity V_g of the drops has reached 96% of the velocity u_G of the gas, there is practically no more capture possible by means of inertia.

If, in the above expression, we set $V_g = 0.96\ u_G$ for the upper bound, and if we use the value V_{g1} of the velocity of the drops immediately after the shock between the gas and the liquid, we obtain:

$$l_c = k_g \left[\frac{0.616\ u_G^{0.6}}{0.24 \times (0.04)^{0.4}} - \frac{u_G - 0.4V_{g1}}{0.24 \left(u_G - V_{g1}\right)^{0.4}} \right]$$

3.1.14. *Theoretical dust capture yield*

The quenching liquid pulverized at the inlet to the convergent at the *same* velocity as the gas does not stop dust. On the other hand, the initially *immobile* liquid dispersed by the gas at the entrance to the neck, whose flowrate is much higher than that of the quenching liquid, captures all the dust. This liquid, then, is the dust-removal liquid.

We know that the separation equation is:

$$-\frac{dc}{c} = \frac{3\eta_u Q_L}{2d_g V_g A_C} dl$$

The gas, which in the neck is animated with the velocity u_G, takes time $d\tau$ to travel the length $d\ell$:

$$dl = u_G d\tau$$

However, the motion of the drops obeys the equation:

$$d\tau = -k_g \frac{dV_R}{V_R^{1.4}}$$

By eliminating $d\ell$ and $d\tau$ between the last three equations, we obtain:

$$\frac{dc}{c} = \frac{3\eta_u Q_L u_G k_g dV_R}{2 d_g V_g A_C V_R^{1.4}}$$

However:

$$k_g = \frac{4\rho_L d_g}{3 \times 18.5 \times \rho_G} \left[\frac{d_g \rho_G}{\mu_G}\right]^{0.6}$$

Thus:

$$\frac{dc}{c} = \frac{2\rho_L Q_L u_G}{18.5\ \rho_G A_C} \left[\frac{d_g \rho_G}{\mu_G}\right]^{0.6} \frac{\eta_u dV_R}{V_g V_R^{1.4}} \qquad (dV_R < 0)$$

Let us set:

$$K = \frac{2\rho_L Q_L u_G}{18.5\ \rho_G A_C} \left[\frac{d_g \rho_G}{\mu_G}\right]^{0.6}$$

In addition, we know that:

$$\eta_I = \frac{V_R^2}{V_R^2 + V_0^2} \qquad \text{and} \qquad V_g = u_G - V_R$$

By integration:

$$\mathrm{Ln}\frac{c_2}{c_1} = K \int_{\text{inlet}}^{\text{outlet}} \frac{V_R^{0.6} dV_R}{(u_G - V_R)(V_R^2 + V_0^2)}$$

or indeed:

$$-\text{Ln}\frac{c_2}{c_1} = K\int_{\text{inlet}}^{\text{outlet}} \frac{V_R^{0.6} dV_R}{\left(u_G - V_R\right)\left(V_R^2 + V_0^2\right)} = KI$$

The venturi's theoretical capture yield is therefore:

$$\eta_T = 1 - \frac{c_2}{c_1} = 1 - \exp^{-KI}$$

3.1.15. *Capture yield in practice*

The above implies that the liquid drops are uniformly distributed within the gas stream. However, this is not truly the case, because those drops follow preferred trajectories.

Thus, finally, we write:

$$\eta = 1 - \exp^{-RKI}$$

R is a coefficient which expresses the non-uniformity of the distribution of the liquid, and we adopt the hypothesis that:

$$R = 0.1$$

3.1.16. *Integration limits for the integral I*

When the liquid threads and films are injected at the neck, we understand that they encounter a shock with the gas, and that each elementary volume $d\Omega$ of gas encountering the liquid displaces a volume equal to its own, also endowing the liquid with its kinetic energy. At the same time, the liquid divides into drops.

In other words, if V_{g0} is the velocity of the drops after the shock and u_G is the velocity of the gas, we can write:

$$d\Omega \frac{\rho_G u_G^2}{2} = d\Omega \frac{\rho_L V_{g0}^2}{2}$$

That is:

$$V_{g0} = u_G \sqrt{\frac{\rho_G}{\rho_L}}$$

As only a very small portion (around 1%) of the space of the neck is occupied by the liquid, we can assume that the velocity and pressure of the gas remain unchanged.

At the outlet from the neck, the velocity of the drops is practically equal to that of the gas, meaning that the relative velocity $V_R = u_G - V_g$ is zero.

Finally, the relative velocity varies from:

$$V_{R0} = u_G - V_{g0} \quad \text{at the entrance to the neck}$$

to $\quad V_{R1} \# 0 \quad$ at the exit from the neck

and, in view of the approximate expression of the integral 1 given in section 3.6.1, this integral is calculated on the basis of:

$$y_0 = \frac{V_{R0}^{0.6}}{\left(u_G - V_{R0}\right)\left(V_{R0}^2 + V_0^2\right)}$$

$$y_{1/2} = \frac{\left(V_{R0}/2\right)^{0.6}}{\left[u_G - V_{R0}/2\right]\left[\left(V_{R0}/2\right)^2 + V_0^2\right]}$$

when:

$$I = J\left(V_{R0}\right) \quad \text{because} \quad J\left(V_{R1}\right) = 0$$

3.2. Example of simulation of a venturi

3.2.1. *Energy balance at the entrance to the convergent*

$C_V = 1874\ \text{J.kg}^{-1}.\text{K}^{-1}$	$W_{VE} = 0.715\ \text{kg.s}^{-1}$	$M_V = 18$
$C_I = 980\ \text{J.kg}^{-1}.\text{K}^{-1}$	$W_I = 2.618\ \text{kg.s}^{-1}$	$M_I = 31$

$C_L = 4180\ J.kg^{-1}.K^{-1}$ $W_{GE} = 3.333\ kg.s^{-1}$

$T_{GE} = 380°C$ $y_{VE} = 0.32$ (to be verified)

$T_{LE} = 20°C$ $L_0 = 2.5\times10^6\ J.kg^{-1}$

$P_E = 1$ bar abs $W_{LE} = 2.222\ kg.s^{-1}$

$$q_E = 380(2.618\times980 + 0.715\times1874) + 20\times2.222\times4180 + 2.5\times10^6\times0.715$$

$$q_E = 3,457,539$$

$$y_{v0}^{(0)} = \frac{0.715/18}{\frac{0.715}{18} + \frac{2.618}{31}} = 0.32$$

$$\pi^{(0)} = 0.32 \text{ bar abs} \rightarrow T_{M0}^{(0)} = 70.5°C$$

$$E^{(0)} =$$

$$\frac{3,457,539 - 70.5(2.618\times980 + 0.715\times1874 + 2.222\times4180) - 2.5\times10^6\times0.715}{2.5\times10^6 + 70.5(1874 - 4180)}$$

$$E^{(0)} = 0.3165\ kg.s^{-1} > 0$$

$$W_V^{(0)} = 0.715 + 0.3165 = 1.0703\ kg.s^{-1}$$

$$y_{V0}^{(1)} = \frac{1.0703/18}{\frac{1.0703}{18} + \frac{2.618}{31}} = 0.40$$

$$\pi^{(1)} = 0.40 \text{ bar abs } \Delta T_{M0}^{(1)} = 76°C$$

$$E^{(1)} = \frac{3,457,539 - 76\times13,194 - 1,787,500}{2.6\times10^6 - 76\times2306}$$

$$E^{(1)} = 0.318\ kg.s^{-1}$$

$$W_v^{(1)} = 0.715 + 0.318 = 1.033 \text{ kg.s}^{-1}$$

$$y_{V0}^{(2)} = \frac{1.033}{1.033 + 1.52} = 0.404$$

$$\pi^{(2)} = 0.404 \text{ bar abs } \Delta T_{M0}^{(2)} = 76°\text{C} = 349\text{K}$$

$$M_{G0} = 0.404 \times 18 + (1 - 0.404) \times 31 = 25.348 \text{ kg.kmol}^{-1}$$

$$W_{L0} = 2.22 - 0.318 = 1.904 \text{ kg.s}^{-1}$$

$$\rho_{G0} = \frac{25.348 \times 10^5}{8314 \times 349} = 0.8736 \text{ kg.m}^{-3}$$

$$W_{G0} = 2.618 + 1.033 = 3.651 \text{ kg.s}^{-1}$$

$$W_{V0} = 1.033 \text{ kg.s}^{-1}$$

3.2.2. *Pressure at the entrance to the neck*

$$A_E = 0.44 \text{ m}^2 \qquad A_C = 0.06 \text{ m}^2$$

Equation of the impulsion applied to the convergent:

$$\left[P_E^2 - P_C^2\right]^{(0)} = \frac{(3.651 + 1.904)3.651 \times 8314 \times 349}{25.348}\left[\frac{1}{(0.06)^2} - \frac{1}{(0.44)^2}\right] = 6.329 \times 10^8$$

$$P_C^{(0)} = \left[10^{10} - 6.329 \times 10^8\right]^{1/2} = 96783\text{Pa}$$

$$\left[P_E^2 - P_C^2\right]^{(1)} = 23.216 \times 10^5\left[\frac{1}{(0.06)^2} - \frac{1}{(0.044)^2} + \frac{2}{3}\left[\frac{2}{(0.06)^2} + \frac{1}{(0.44)^2}\right]\text{Ln}\frac{1}{0.96827}\right]$$
$$= 6.612 \times 10^8$$

$$P_C^{(1)} = \left[10^{10} - 6.612 \times 10^8\right]^{1/2} = 96{,}637 \text{ Pa}$$

3.2.3. *Heat balance at the entrance to the neck*

$$y_{V1}^{(0)} = 0.397 \left(\text{estimated; we could have taken: } y_{V1}^{(0)} = y_{VE} = 0.32\right)$$

$$\pi = 0.397 \times 0.96637 = 0.384$$

$$T_{M1}^{(0)} = 75°C \qquad q_E = 3{,}457{,}539 \text{ J.s}^{-1}$$

$$E =$$

$$\frac{3{,}457{,}539 - 75(2.618 \times 980 + 1.033 \times 1874 + 1.904 \times 4180) - 2.5 \times 10^6 \times 1.033}{2.5 \times 10^6 + 75(1874 - 4180)}$$

$$E^{(0)} = -0.025 \text{ kg.s}^{-1}$$

$$W_{V1} = 1.033 - 0.025 = 1.008 \text{ kg.s}^{-1}$$

$$y_{V1}^{(1)} = \frac{1.008}{1.008 + 1.52} = 0.398$$

$$\pi = 0.398 \times 0.96637 = 0.385 \text{ bar}$$

$$T_{M1}^{(1)} = 75°C$$

From this, we obtain the conditions at the entrance to the neck:

$$W_{LI} = 1.904 + 0.025 = 1.929 \text{ kg.s}^{-1}$$

$$W_{VI} = 1.008 \text{ kg.s}^{-1}$$

$$W_I = 2.618 \text{ kg.s}^{-1}$$

$$W_{GI} = 1.008 + 2.618 = 3.626 \text{ kg.s}^{-1}$$

$$T_{MI} = 75°C = 348 \text{ K}$$

$$M_{GI} = 0.398 \times 18 + (1 - 0.398) \times 31 = 25.766 \text{ kg.kmol}^{-1}$$

$$\rho_{GI} = \frac{25.766 \times 96{,}637}{8314 \times 348} = 0.8618 \text{ kg.m}^{-3}$$

We suppose that the temperature of the scrubbing water injected at the entrance to the neck is equal to that of the gas. Indeed, this water is recirculated from the outlet of the venturi. As the gas pressure has decreased slightly, its temperature decreases a little as well, as it remains saturated.

3.2.4. *Pressure drop in the neck*

$$W_{LD} = 5.555 \text{ kg.s}^{-1}$$

$$Q_{GI} = \frac{3.626}{0.8618} = 4.2 \text{ m}^3.\text{s}^{-1}$$

$$u_{GI} = \frac{4.2}{0.06} = 70 \text{ m.s}^{-1}$$

$$\Delta P_{EL} = \frac{5.555 \times 70}{0.06} = 6,481 \text{ Pa}$$

This gives us the pressure at the exit from the neck:

$$P_{SC} = 96,637 - 6,481 = 90,156 \text{ Pa}$$

As the temperature is unchanged, the density becomes:

$$\rho_{G2} = \frac{25.766 \times 90,156}{8,314 \times 348} = 0.8029 \text{ kg.m}^{-3}$$

$$Q_{G2} = 3.626/0.8029 = 4.516 \text{ m}^3.\text{s}^{-1}$$

The mean pressure in the neck is:

$$\overline{P_C} = 96,637 - \left(\frac{6,481}{2}\right) = 93,396 \text{ Pa}$$

and the mean density in the neck is:

$$\overline{\rho}_{GC} = \frac{25.766 \times 93,396}{8,314 \times 348} = 0.8317 \text{ kg.m}^{-3}$$

3.2.5. *Pressure at the outlet from the divergent*

$$\Delta\left(P^2\right)=\left[0.6\times3.62+0.1\times\left(1.929+5.555\right)\right]\frac{3.62\times8,314\times348}{25.766}\left[\frac{1}{(0.06)^2}-\frac{1}{(0.3)^2}\right]$$

$$\left(P^2\right)=3.175\times10^8\,\text{Pa}^2$$

$$P_S=\left[90,145^2+3.175\times10^8\right]^{1/2}$$

$$P_S=91,889\text{ Pa}$$

Hence, we have the overall pressure drop in the device:

$$10^5-91,889=8,111\text{ Pa}$$

3.2.6. *Diameter of the drops of scrubbing water*

$$d_g=\frac{0.004964}{70}+0.0288\left[\frac{0.0055}{4.2}\right]^{1.5}$$

$$d_g=72\times10^{-6}\text{m}$$

3.2.7. *Length of the neck*

$$k_g=\frac{4\times1,000\times10^{-6}}{3\times18.5\times0.8317}=0.00623$$

This tells us the length of the neck:

$$l_c=0.00623\left[\frac{0.616\times70^{0.6}}{0.24\times0.276}-\frac{70-0.4\times2.5}{0.24(70-2.5)^{0.4}}\right]$$

$$l_c=0.4\text{ m}$$

3.2.8. *Capture yield*

$d_P = 10^{-6}$ m $\rho_S = 1{,}500$ kg.m^{-3}

$\mu_G = 10 \times 10^{-6}$ Pa.s $\rho_L = 1{,}000$ kg.m^{-3}

$$K = \frac{2 \times 5.555 \times 70}{18.5 \times 0.8317 \times 0.06}\left[\frac{72 \times 10^{-6} \times 0.8317}{20 \times 10^{-6}}\right]^{0.6}$$

$$K = 1{,}626$$

$$V_{gI} = 70\sqrt{\frac{0.8618}{1{,}000}} = 2.06 \text{ m.s}^{-1}$$

$$V_{RI} = 70 - 2.06 = 68 \text{ m.s}^{-1}$$

$$V_0 = \frac{20.6 \times 72 \times 10^{-6} \times 20 \times 10^{-6}}{1{,}500 \times 10^{-12}} = 19.78 \text{ m.s}^{-1}$$

$$V_0^2 = 391 \text{ m}^2.\text{s}^{-2}$$

$$y_0 = \frac{68^{0.6}}{2.06\left(68^2 + 391\right)} = 1.215 \times 10^{-3}$$

$$y_{1/2} = \frac{34^{0.6}}{36\left(34^2 + 391\right)} = 0.148 \times 10^{-3}$$

$$I = \frac{68}{3}\left[\frac{1.215}{2} + 0.296\right]10^{-3} = 0.0205$$

$$R = 0.1$$

$$\eta = 1 - e^{-1626 \times 0.0205 \times 0.1}$$

$$\eta = 0.965$$

The true value is somewhat higher than this, because we did not take account of the condensation following recompression in the divergent. In addition, the calculations did not take account of the granulometric spread of the dust and the drops.

3.3. Pulverization columns

3.3.1. *Theoretical capture yield*

We know that the separation equation is:

$$-\frac{dc}{c}=\frac{3\eta_u Q_L}{2d_g V_g A}dH = K\eta_u Q_L dH$$

H: height of the column (m).

V_g is the absolute velocity of the drops as they fall – i.e. their limiting velocity V_ℓ of fall in the gas, less the gas's own velocity of ascension in an empty bed.

$$\rho_L \frac{\pi}{6} d_g^3 g = C_x \frac{\pi d_g^2}{4} \frac{\rho_G V_l^2}{2}$$

When the size of the drops is between 100 and 500µm:

$$C_x = \frac{18.5}{R_e^{0.6}} = \frac{18.5}{V_l^{0.6}}\left[\frac{\mu_G}{d_g \rho_G}\right]^{0.6}$$

By eliminating C_x between these two equations:

$$V_l = \left[\frac{4 d_g \rho_L g}{55.5_G}\left[\frac{d_g \rho_G}{\mu_G}\right]^{0.6}\right]^{0.7143}$$

η_u is the unitary yield of the drops.

$$\eta_u = \frac{V_R^2}{V_R^2 + V_0^2}$$

V_g is the absolute velocity of the drops – i.e. their velocity in relation to the outside world. At the output from the pulverizer, this velocity V_{g0} is of the order of 3m.s^{-1}. At the foot of the column, the velocity V_g is reduced to:

$$V_{g1} = V_1 - u_G = V_1 - 0.6\ V_1 = 0.4\ V_1 \quad \text{(justified hereafter)}$$

The velocity u_G is the gas's absolute velocity in an empty bed.

We shall accept the hypothesis that:

$$V_g = \frac{1}{3}\left(V_{g0} + 2V_{g1}\right)$$

V_R is the relative velocity of the drops in relation to the gas. At the output from the pulverizer, we have:

$$V_{R0} = V_{g0} + u_G$$

At the bottom of the column:

$$V_{R1} = V_1$$

We shall accept that:

$$V_R = \frac{1}{3}\left(V_{R0} + 2V_{R1}\right)$$

The constant V_0 is given by:

$$V_0 = \frac{20.6\ d_g \mu_G}{\Delta\rho d_p^2}$$

The absolute velocity of the gas in an empty bed must be very limited if we do not wish to see significant entrainment of liquid droplets. Indeed, it is impossible to install a droplet-separating mat, as it would quickly clog. Only a cyclone can be used.

The simplest solution is to accept the following, for the velocity of the gas in an empty bed:

$$u_G = 0.6V_l$$

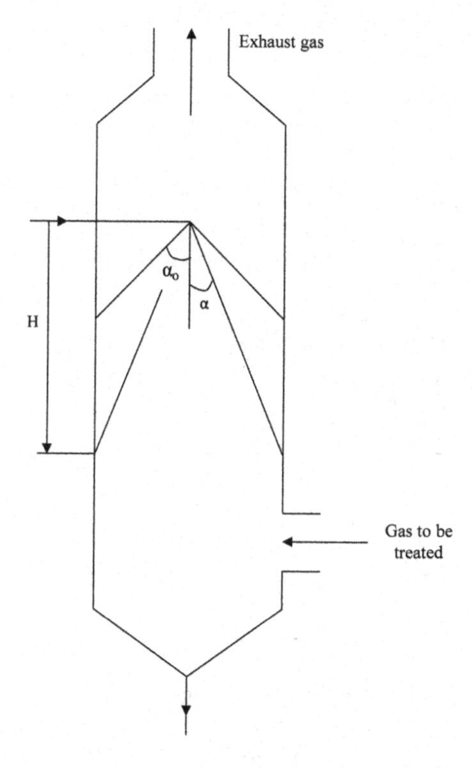

Figure 3.2. *Pulverization column*

Now let α_o be the half-angle at the summit of the pulverization cone. We shall suppose that this is a full cone. The relation between the solid angle Ω and α is:

$$\Omega = 2\pi(1 - \cos\alpha)$$

The useful flowrate of liquid at the height H is:

$$Q_{LU} = Q_{L0} \times \frac{\Omega}{\Omega_0} = Q_{L0} \frac{1 - \cos\alpha}{1 - \cos\alpha_o}$$

However, the separation equation is written with H counted positively from top to bottom:

$$-\frac{dc}{c} = K\eta_u Q_{LU} dH = \frac{K\eta_u Q_{L0}}{1-\cos\alpha_o}(1-\cos\alpha)dH$$

where:

$$K = \frac{3}{2d_g V_g A}$$

$$\text{Ln}\frac{c}{c_0} = -\frac{K\,Q_{L0}}{1-\cos\alpha_o}\left[H - H_0 - \int_{\alpha_o}^{\alpha}\cos\alpha dH\right] \qquad [3.1]$$

However, if D is the diameter of the column:

$$H = \frac{D}{2tg\alpha} = \frac{D\cos\alpha}{2\sin\alpha} \quad \text{and} \quad dH = -\frac{D}{2}\frac{1}{\sin^2\alpha}d\alpha$$

Therefore:

$$\int_{\alpha_o}^{\alpha}\frac{\cos\alpha}{\sin^2\alpha}d\alpha = \frac{1}{\sin\alpha_o} - \frac{1}{\sin\alpha}$$

The square bracket in equation [3.1] becomes:

$$\frac{D}{2}\left[\frac{1}{tg\alpha} - \frac{1}{tg\alpha_o} - \frac{1}{\sin\alpha} + \frac{1}{\sin\alpha_o}\right] = \frac{D}{2}R$$

where:

$$R = \frac{1-\cos\alpha_o}{\sin\alpha_o} - \frac{1-\cos\alpha}{\sin\alpha}$$

Thus, after integration:

$$\text{Ln}\frac{c}{c_o} = -\frac{K\eta_u Q_{L0} DR}{2(1-\cos\alpha_o)}$$

The capture yield is:

$$\eta = 1 - \frac{c}{c_o} = 1 - \exp\left[-\frac{K\eta_u Q_{L0} DR}{2(1-\cos\alpha_o)}\right]$$

3.3.2. *Pressure drop of the gas in the column*

When drops of water fall, the work due to gravity is dissipated in friction against the gas, in the proportion of around 99%, and the remaining ~1% is used to give the drops their kinetic energy. We shall discount this second term, which pertains only to the pulverizer.

For the sake of safety and simplicity, we shall also ignore that fact that a portion of the liquid flows along the walls and, therefore, no longer contributes to slowing down the gas.

Finally, if W_L is the liquid flowrate, we can write that the power developed by gravity over the height H balances out the pressure drop ΔP_G of the gaseous flowrate Q_G:

$$W_L gH = Q_G \Delta P_G$$

This gives us ΔP_G.

EXAMPLE 3.1.–

$d_g = 100\times10^{-6}$ m　　$\rho_L = 1{,}000$ kg.m^{-3}

$H = 8$ m　　$\rho_G = 0.9$ kg.m^{-3}

$u_G = 0.6\ V_\ell$　　$\mu_G = 20\times10^{-6}$ Pa.s

$\alpha_o = 30°$　$\cos\alpha_o = 0.866$　　$D = 1.5$ m

$\rho_S = 1{,}500$ kg.m^{-3}　　$W_L/W_G = 8$

$V_{g0} = 3$ m.s^{-1}　　$d_p = 1.79\times10^{-6}$ m

$$V_1 = \left[\frac{4\times10^{-4} \times 1,000 \times 9.81}{55.6\times0.9} \left[\frac{100\times10^{-6}\times0.9}{20\times10^{-6}} \right]^{0.6} \right]^{0.7163}$$

$$V_1 = 0.308\ \text{m.s}^{-1}$$

$$V_0 = \frac{20.6\times100\times10^{-6}\times20\times10^{-6}}{1,500\times\left(1.79\times10^{-6}\right)^2}$$

$$V_0 = 8.608 \quad \text{and} \quad V_0^2 = 74\ \text{m}^2.\text{s}^{-2}$$

$$V_{g1} = 0.4\times0.308 = 0.123\ \text{m.s}^{-1}$$

$$V_{g0} = 3\ \text{m.s}^{-1}$$

$$V_g = \frac{1}{3}(0.246+3) = 1.08\ \text{m.s}^{-1}$$

$$u_G = 0.6\times0.308 = 0.185$$

$$V_{R0} = 3+0.185 = 3.185$$

$$V_{R1} = 0.308\ \text{m.s}^{-1}$$

$$V_R = \frac{1}{3}(3.185+0.616) = 1.27\ \text{m.s}^{-1}$$

$$\eta_u = \frac{1.27^2}{1.27^2+75.613} = 0.021$$

$$A = \frac{\pi}{4}\times1.5^2 = 1.766\ \text{m}^2$$

$$K = \frac{3}{2\times100\times10^{-6}\times1.08\times1.766} = 7859$$

$$\alpha_o = 30° \qquad \cos\alpha_o = 0.866 \qquad \sin\alpha_o = 0.5$$

$$\alpha = \text{Arc tg}\left[\frac{D}{2H}\right]_o = \text{Arc tg}0.0937$$

$$\alpha = 5.356 \text{ degrees}$$

$$\cos\alpha = 0.9956 \qquad \sin\alpha = 0.0933$$

$$R = \frac{1-0.866}{0.5} - \frac{1-0.9956}{0.0933} = 0.02208$$

$$W_G = 1.766 \times 0.6 \times 0.308 \times 0.9 = 0.293 \text{ kg.s}^{-1}$$

$$W_L = 8 \times 0.293 = 2.344 \text{ kg.s}^{-1}$$

$$Q_{LO} = 0.002344 \text{ m}^3.\text{s}^{-1}$$

$$-\text{Ln}\frac{c}{c_o} = \frac{7,859 \times 0.021 \times 0.002344 \times 1.5 \times 0.2208}{2(1-0.866)}$$

$$-\text{Ln}\frac{c}{c_o} = 0.478$$

$$\eta = 0.38$$

The pressure drop in the column is:

$$\Delta P_G = \frac{2.344 \times 9.81 \times 8}{0.293} = 565 \text{ Pa}$$

NOTE.–

As we know from the above method, Ln (c/c_o) varies closely with the fourth power of the diameter of the dust particles. Thus, there is a clear divide between dust that is stopped completely and that which is not stopped at all.

Compared to a venturi, these columns are much less costly in terms of initial investment, but their yield when dealing with fine particles (of around a µm) is mediocre.

3.4. Various points

3.4.1. *Other types of wet scrubbers*

3.4.1.1. *Bubbling devices*

These devices are capable of handling high dust contents, and can be used for rough milling. They consume only a small amount of water (e.g. 0.05 liters per m^3 of gas), and the pressure drop of the gas is only around 1600–2000 Pa.

Ultimately, their performances are intermediary between those of the pulverization column and the venturi.

3.4.1.2. *Perforated plates*

These plates are inspired by the plates used in distillation, but the velocity passing through the holes, in the present case, is around three times greater. Thus, in the vicinity of atmospheric pressure, we accept a velocity of around 25 $m.s^{-1}$. With certain manufacturers, the gas jet is flattened against a deflector placed in front of each hole.

The pressure drop on passing through the holes is obviously not insignificant, owing to the high velocity, but the high value of the gas velocity slows down or even eliminates the clogging of the holes.

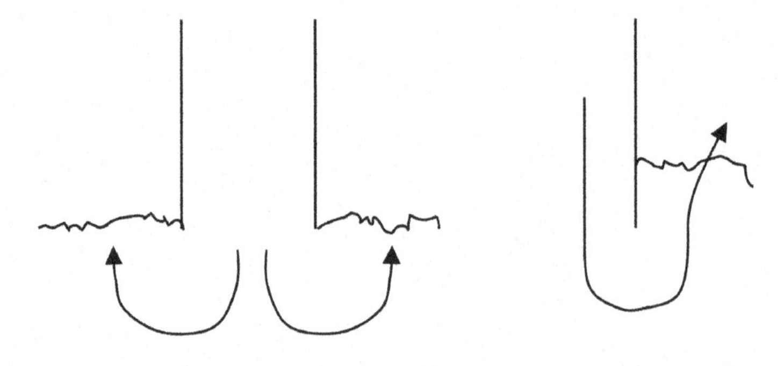

Gas jet incident to the surface of the liquid

Partially-submerged chicane

Figure 3.3. *Bubbling devices*

3.4.2. *General conclusion on wet scrubbing*

This type of dust removal exhibits the following advantages:

– a very hot gas can be cooled;

– the dangers of explosion are eliminated, because the relative humidity is 100% and, in these conditions, there can be no occurrence of electrostatic charges and sparks;

– these devices are compact (apart from pulverization columns or perforated-plate columns).

However, certain precautions must be exercised:

– we must guard against freezing;

– we must provide an adequate protective coating if the gas contains corrosive components.

3.5. Choice of an air scrubber

In his Table II, Stairmand [STA 65] gives the capture yield for 21 different types of scrubbers for three particle sizes: 1 μm, 5 μm and 50 μm. His Figure 32 gives an idea of the cost (investment and running costs) of these devices. Finally, he touches on the problem of the dispersion of a cloud of fumes in the atmosphere.

Hansberg [HAN 61] gives practical data on wet scrubbers. Pozin *et al.* [POZ 57] also provide practical data concerning perforated-plate scrubbers. Finally, Taheri and Calvert [TAH 68] present information on foam-plate columns.

More specifically, Calvert [CAL 77a], Semrav [SEM 77] and Mottola [MOT 77] consider the operation of wet scrubbers. For his part, Calvert [CAL 77b] proposes ways to improve the operation of wet scrubbers.

NOTE.–

In section 3.6.2, readers will find the overall heat balance for a wet scrubber.

3.6. Varied calculations

3.6.1. *Approximate calculation of the integral*

We can assimilate the integrand to a third-degree polynomial:

$$f(V_R) = y = aV_R^3 + bV_R^2 + cV_R$$

Indeed, the constant term has a zero value, because the integrand becomes zero when $V_R = 0$. We need to integrate from V_{R0} to V_{R1}.

In addition, we shall accept the hypothesis that:

– the value of $f(V_R)$ is exact for $x = V_{RO}$ and $x = V_{RO}/2$

– the value of dy/dx is exact for $V_R = V_{R0}$.

Thus:

$$y_0 = aV_{R0}^3 + bV_{R0}^2 + cV_{R0}$$

$$y_{1/2} = \frac{a}{8}V_{R0}^3 + \frac{b}{4}V_{R0}^2 + \frac{C}{2}V_{R0}$$

$$y'_0 = 3aV_{R0}^2 + 2bV_{R0} + c$$

Hence:

$$a = \frac{-6y_0 + 8y_{1/2} + 2V_{R0}y'_0}{V_{R0}^3}$$

$$b = \frac{11y_0 - 16y_{1/2} - 3V_{R0}y'_0}{V_{R0}^2}$$

$$c = \frac{1}{V_{R0}}\left[-4y_0 + 8y_{1/2} + V_{R0}y'_0\right]$$

However:

$$I = \int_{V_{R1}}^{V_{R0}} \left(aV_R^3 + bV_R^2 + cV_R\right) dV_R = \left[\frac{aV_R^4}{4} + \frac{bV_R^3}{3} + \frac{cV_R^2}{2}\right]_{V_{R1}}^{V_{R0}}$$

If we replace a, b and c with their values and set $V_R/V_{R0} = x$, the primitive of f(x) is written:

$$J(V_R) = V_R y_0 \left(\frac{-3}{2}x^3 + \frac{11}{3}x^2 - 2x\right) + V_R y_{1/2}\left(2x^3 - \frac{16}{3}x^2 + 4x\right)$$
$$+ V_R^2 y'_0 \left(\frac{1}{2}x^2 - x + \frac{1}{2}\right)$$

and, if $V_R/V_{R0} = x = 1$,

$$J(V_{R0}) = \frac{V_{R0}}{3}\left[\frac{y_0}{2} + 2y_{1/2}\right]$$

Finally:

$$I = J(V_{R0}) - J(V_{R1})$$

3.6.2. *Calculation of the overall heat balance for a wet scrubber*

Fumes at T_E, y_{VE} and Q_{GE} ($Nm^3.s^{-1}$), and P_E:

vapor: $y_{VE} = Q_{VE}/Q_{GE}$

inerts: $y_{IE} = 1 - y_{VE}$

Watering rate for quenching (cooling + saturation):

$$A = W_L/Q_{GE}$$

Heat extracted per Nm^3 of fumes:

$$q_E = T_E C_G + AC_L t_L = AC_L t_L + T_E\left(C_I y_{IE} + C_V y_{VE}\right)$$

We make a hypothesis concerning the output temperature $T_S^{(0)}$ common to both the water and the fumes. This gives us the vapor tension at T_S:

$$\pi_S = \pi(T_S)$$

If P_S is the total pressure at the outlet, the water vapor flowrate at the outlet is:

$$Q_{VS} = Q_{GE} y_{IE} \frac{\pi_S}{P_S - \pi_S}$$

Let us set:

$$\Psi = \frac{Q_{VS} - Q_{VE}}{Q_{GE}} = \frac{y_{IE}\pi_S}{P_S - \pi_S} - y_{VE}$$

If:

$$\Psi > 0 \rightarrow \text{vaporization}$$

$$\Psi < 0 \rightarrow \text{condensation}$$

If ρ_{VN} is the density of the water vapor in normal conditions:

$$\rho_{VN} = 18/22.42 = 0.8028 \text{ kg.Nm}^{-3} \quad (\text{for water})$$

Let us set:

$$\phi = \rho_{VN}\Psi$$

The heat balance must be satisfied:

$$q_E = T_S(C_G + C_L A) + \phi L_S$$

where:

$$L_S = H_S - h_S$$

L_S is the latent heat of vaporization at T_S.

Hence:

$$T_S^{(1)} = \frac{q_E - \phi L_S}{C_G + C_L A}$$

We then merely need to repeat the calculation of $\Psi^{(1)}$, $\phi^{(1)}$ and $T_S^{(2)}$, etc.

EXAMPLE 3.2.–

$C_L = 4{,}180\ J.kg^{-1}.K^{-1}$ $t_L = 20°C$ $P_S = 0.9$ bar

$C_G = 1{,}450\ J.(Nm^3.K)^{-1}$ $T_E = 350°C$ $\rho_{VN} = 0.8028$

$y_{VE} = 0.32$ $A = 0.8\ kg.Nm^{-3}$

$$q_E = 350 \times 1{,}450 + 4{,}180 \times 0.8 \times 20$$

$$q_E = 5.075 \times 10^5 + 0.6688 \times 10^5 = 5.7438 \times 10^5$$

$$T_S^{(0)} = 73°C \quad (\text{estimated})$$

$$\pi_S = 0.3546 \text{ bar} \qquad L_S = 2.321 \times 10^6\ J.kg^{-1}$$

$$\Psi^{(0)} = \frac{0.68 \times 0.3546}{0.9 - 0.3546} - 0.32 = 0.1221$$

$$\phi^{(0)} = 0.1221 \times 0.8028 = 0.09803$$

$$T_S = \frac{5.7438 \times 10^5 - 0.09803 \times 2.321 \times 10^6}{4{,}814} = 71.6°C$$

$$T_S^{(1)} = 72.9°C \,(\text{value close to } 73°C)$$

$$\pi_S = 0.3531 \text{ bar} \qquad L_S = 2.326 \times 10^6\ J.kg^{-1}$$

$$\Psi^{(1)} = \frac{0.68 \times 0.3531}{0.9 - 0.3531} - 0.32 = 0.1190$$

$$\phi^{(1)} = 0.095561$$

$$T_S^{(2)} = \frac{5.7438 \times 10^5 - 0.095561 \times 2.326 \times 10^6}{4,844} = 72.6°C$$

T_S is very slightly less than 72.9°C.

4

Separation Between a Fluid and a Divided Solid Through Centrifugal Force

4.1. The cyclone

4.1.1. *Principle and precautions for use*

The fluid (either a gas or a liquid) enters tangentially into a cylindrical chamber, and is thus endowed with a helical motion during which centrifugal force drives the solid particles in suspension onto the vessel walls. Ultimately, this is forced decantation.

An important parameter is the velocity V_e of the fluid on entry:

– for gases whose pressure is near to 1 atmosphere, this velocity must be no less than 15 $m.s^{-1}$ for a sufficient yield and no greater than 20–25 $m.s^{-1}$ so as not to re-entrain the deposited solid;

– for liquids, the velocity at the inlet must be no less than 1.5$m.s^{-1}$ for an acceptable yield and that velocity must be no greater than 2 $m.s^{-1}$ (between 2 and 3 $m.s^{-1}$, the cyclone must be made from an abrasion-resistant material if the solid phase is hard).

4.1.2. *Phases and components*

The proportions often adopted are those given by Zenz [ZEN 75]. They are illustrated in Figure 4.1. For hydrocyclones, Rietema [RIE 61] proposed different proportions, but then the device is more difficult to create and, ultimately, the proportions below ensure satisfactory yield and operation.

Width of the inlet: $L_e = D_c/4$

Height of the inlet: $H_e = D_c/2$

Diameter of the chimney: $D_i = D_c/2$

Draft of the chimney: $S_i = D_c/8$

Cylindrical height: $H_{cy} = 2\ D_c$

Conical height: $H_{co} = 2\ D_c$

Total height of cyclone: $h = 4\ D_c$

Diameter of underflow: $D_s = D_c/4$

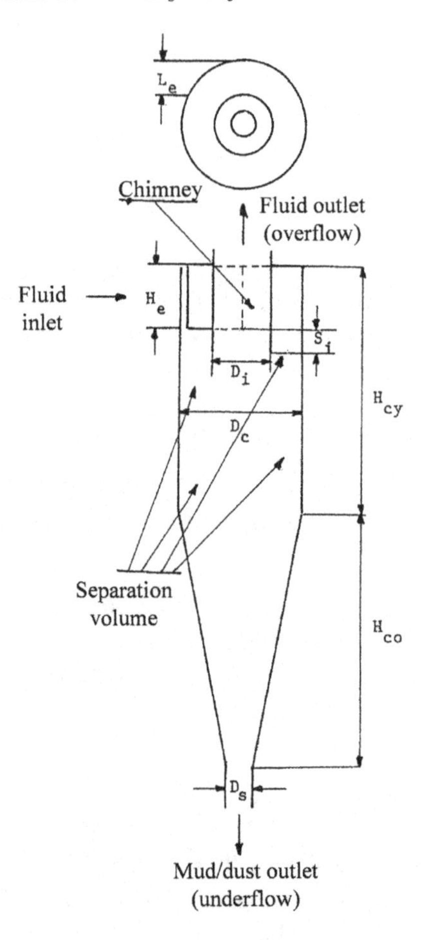

Figure 4.1. *Proportions of a cyclone*

We can see that in the immediate vicinity of the cyclone, the inlet has a rectangular cross-section.

4.1.3. *Presentation of the pressure drop*

By expounding some of the arguments put forward by Barth, Muschelknautz and Brunner [BAR 56, MUS 72, BRU 80], we shall show that this loss in pressure is attributable to:

– the loss of kinetic moment by the fluid because of friction in the separation volume;

– the dissipation of energy to work against the centrifugal force when the mixture returns to the inlet of the chimney.

4.1.4. *Pressure drop in the separation volume*

The kinetic torque in relation to the axis of the cyclone of the fluid on *entry* into the separation volume is:

$$M_E = \rho_F Q_F r_c U_m = \rho_F Q_F r_m U_E$$

This equation defines U_E.

ρ_F: density of the fluid (kg.m^{-3})

Q_F: volumetric fluid flowrate (m^3.s^{-1})

r_c: radius of the cyclone ($r_c = D_c/2$) (m)

U_m: mean orthoradial velocity of the fluid (m.s^{-1})

r_m: mean radial distance of the fluid (m.s^{-1})

Similarly, at the *outlet* from the separation volume, and just before returning to the inlet of the chimney, the kinetic torque of the fluid is:

$$M_i = \rho_F Q_F r_i U_i = \rho_F Q_F r_m U_S$$

This equation defines U_S.

r_i: radius of the chimney ($r_i = D_i/2$) (m)

U_i: orthoradial velocity of the fluid immediately before entering the chimney ($m.s^{-1}$)

The velocities U_F and U_S express the kinetic torque with the same lever arm $r_m = \sqrt{r_i r_c}$. The pressure drop in the separation volume is then written:

$$\Delta P_S = \frac{\rho_F}{2}\left(U_E^2 - U_S^2\right) = \frac{\rho_F}{2}\left[U_m^2 \frac{r_c}{r_i} - U_i^2 \frac{r_i}{r_c}\right]$$

Let us now calculate the coefficient ε_S thus defined:

$$\varepsilon_S = \frac{\Delta P_S}{\rho_F \frac{U_i^2}{2}} = \frac{r_i}{r_c}\left[\left[\frac{U_m r_c}{U_i r_i}\right]^2 - 1\right]$$

We replace the ratio ($U_m\, r_c/U_i\, r_i$) with its value given by equation [4.2]. We obtain:

$$\varepsilon_S = \frac{r_i}{r_c}\left[\frac{1}{\left[1 - \frac{U_i h}{V_i r_i}\left(\frac{f}{2} + Bf_s\right)\right]^2} - 1\right]$$

The values U_i/V_i, f, B and f_s are defined in section 4.1.5 by Muschelknautz [MUS 72] and Barth [BAR 56].

4.1.5. *Calculation of U_i/V_i (Barth's method, 1956)*

Let us establish the definitions for the following velocities:

U_i: orthoradial velocity of the fluid at the base of the chimney immediately before entering it ($m.s^{-1}$)

V_i: velocity in an empty bed of the fluid in the chimney ($m.s^{-1}$)

U_m: mean value of the orthoradial velocity in the separation volume ($m.s^{-1}$)

U_c: orthoradial velocity for the calculation of M_S and M_F ($m.s^{-1}$)

Let us also define the kinetic moments of the fluid in relation to the axis of the cyclone:

M_E: torque of the fluid at the top of the separation volume (N.m)

M_i: torque of the fluid immediately before entering the chimney (N.m)

M_F: resistant torque due to friction, as though the solid were absent (fluid alone) (N.m)

M_S: resistant torque due to friction of the fluid against the solid (N.m)

The decrease in the torque of the fluid between the top of the cyclone and the entrance into the chimney is attributable to the friction of the fluid *without* solid and to the friction of the fluid *on* the solid. Thus, we can write:

$$M_E = M_i + M_F + M_S$$

Making various hypotheses regarding to mean values and what could be called the effective values, the authors state the above equation as follows:

$$\rho_F Q_F r_c U_m = \rho_F Q_F r_i U_i + \frac{f}{2} \pi h r_i r_c \rho_F U_i U_m + f_s \pi h r_i r_c \rho_F U_i U_m B \qquad [4.1]$$

f: friction coefficient of the fluid (considered in the absence of solid) on the walls of the cyclone:

$$f = 0.01$$

f_s: friction coefficient of the fluid on the solid at the wall:

$$f_s = 0.25$$

Let us define this last term more precisely. Adopting the hypothesis that the solid moves in a helix on the wall in the form of a single stream with semicircular cross-section profile, Muschelknautz *et al.* [MUS 72] show that the surface S_s of friction of the fluid against the solid is of the form:

$$S_S = Khr_i\sqrt{\eta\mu}\left[\frac{\rho_F}{\rho_a}Fr\right]^{1/2}\left[\frac{r_i}{r_c}\right]^{1/8} = Kh\sqrt{r_i r_c}B$$

where:

$$B = \sqrt{\eta\mu}\left[Fr\frac{\rho_F}{\rho_a}\right]^{1/2}\left[\frac{r_i}{r_c}\right]^{5/8}$$

η: overall capture yield for the cyclone

μ: gas loading ratio

ρ_a: apparent density of the solid deposited at the wall (kg.m^{-3})

The Froude number is:

$$Fr = \frac{V_i}{\sqrt{2gr_i}}$$

g: acceleration due to gravity (9.81 m.s^{-2})

However, M_S is of the form:

$$M_S = r_m\frac{\lambda}{4}S_S\rho_F U_c^2 = \frac{\lambda}{4}\sqrt{r_i r_c}S_S\rho_F U_i U_m$$

Thus, we find the expression of the last term in equation [4.1], of the torques. Let us divide both sides of that equation [4.1] by:

$$r_c U_m \rho_F Q_F = r_c U_m \rho_F \pi r_i^2 V_i$$

We obtain:

$$1 = \frac{r_i U_i}{r_c U_m} + \left(\frac{f}{2} + Bf_s\right)\frac{h}{r_i}.\frac{U_i}{V_i} \qquad [4.2]$$

However, immediately after the entrance, a sudden expansion is inevitable. The velocity is thus diminished, and shifts from the value V_e to the value $U_m = V_e/1.3$. Hence:

$$U_m = \frac{V_e}{1.3} = \frac{\pi r_i^2 V_i}{1.3 A_e}$$

By replacing U_m with that expression, we obtain the result we seek:

$$\frac{U_i}{V_i} = \frac{1}{\dfrac{1.3 A_e}{\pi r_i r_c} + \dfrac{h}{r_i}\left(\dfrac{f}{2} + f_s B\right)}$$

4.1.6. *Energy loss from the fluid upon return*

In the chimney, the fluid at pressure P_h is animated with a helical motion (though it does not occupy the whole cross-section of the chimney). The velocity V_h of the fluid filaments is inclined along the axis of the cyclone. On exiting the chimney, the kinetic energy corresponding to that velocity V_h is converted into pressure, and the fluid is then at the outlet pressure P_s:

$$P_s = P_h + \rho_F \frac{V_h^2}{2}$$

To enter the chimney, the fluid must approach the axis of the cyclone, and thus travel the radial distance Δr, whilst working against the centrifugal force $\rho_F = \frac{U_i^2}{r_i}$. This results in the following pressure drop between the outside of the chimney and the outlet from the cyclone:

$$P_i - P_s = \rho_F \frac{U_i^2}{r_i} \Delta r \qquad [4.3]$$

Indeed, the orthoradial velocity U_i is conserved when the fluid enters the chimney (or may even increase, as we shall see). As, at the entrance to the chimney, the kinetic energy is practically represented by $\rho_F = U_i^2/2$, only the pressure term P_i makes a contribution to equation [4.3].

To penetrate the chimney, Barth [BAR 56] supposes that the fluid undergoes a radial displacement whose amplitude is Δr and, once inside the chimney, it propagates in helical form, occupying an annular space of thickness Δr and limited externally by the wall of the chimney. In that space, the axial component of the oblique velocity V_h is V_a, whereas the axial velocity calculated across the whole cross-section of the chimney is V_i. The equation of conservation of matter is written:

$$2\pi r_i \Delta r V_a = \pi r_i^2 V_i \qquad [4.4]$$

In addition, the different calculations and hypotheses carried out by Barth are tantamount to accepting that the energy consumed against the centrifugal force is used as follows:

– half to create heat and increase the orthoradial kinetic energy;

– the other half to create a kinetic energy corresponding to the axial velocity V_a. It is this energy which is of interest to us and, for the unit volume of the fluid, we can write:

$$\rho_F \frac{V_a^2}{2} = \frac{1}{2}\rho_F \frac{U_i^2}{r_i} \Delta r \qquad [4.5]$$

By eliminating V_a between equations [4.4] and [4.5], we obtain:

$$\frac{\Delta r}{r_i} = \left[\frac{V_i}{U_i}\right]^{2/3} \frac{1}{4^{1/3}} \qquad [4.6]$$

The power lost by the fluid flowrate (in volume) Q_F is then $Q_F\ \Delta P_R$, where:

$$\Delta P_R = P_i + \rho_F \frac{U_i^2}{2} - \left[P_h + \rho_F \frac{V_h^2}{2}\right] = \rho_F \frac{U_i^2}{2} + P_i - P_a$$

Let us now calculate ε_R, defined thus:

$$\varepsilon_R = \frac{\Delta P_R}{\rho_F \frac{U_i^2}{2}} = 1 + \frac{P_i - P_s}{\rho_F \frac{U_i^2}{2}}$$

Let us replace $(P_i - P_s)$ with its expression [4.3], and the ratio $\Delta r/r_i$ with its expression [4.6]. We obtain:

$$\varepsilon_R = 1 + \frac{2/4^{1/3}}{(u_i/V_i)^{2/3}}$$

4.1.7. *Overall pressure drop*

The overall pressure drop is:

$$\Delta P_G = \Delta P_S + \Delta P_R = (\varepsilon_S + \varepsilon_R)\rho_F \frac{U_i^2}{2}$$

However:

$$U_i = \left[\frac{U_i}{V_i}\right] V_i = \left[\frac{U_i}{V_i}\right]\left[\frac{Q_F}{A_i}\right]$$

Thus:

$$\Delta P_G = \frac{\rho_F}{2}(\varepsilon_S + \varepsilon_R)\left[\frac{U_i}{V_i}\right]^2 \left[\frac{Q_F}{A_i}\right]^2$$

NOTE.–

Barth's method [BAR 56], which we have just described, is generally used for gas cyclones. However there is nothing to prevent its being used for hydrocyclones, and Muschelknautz proposes values of f/2 for this application.

EXAMPLE 4.1.–

Gas cyclone:

$Q_F = 0.5\ m^3.s^{-1}$ $\quad$ $\mu = 0.03$

$\rho_F = 1.5\ kg.m^{-3}$ $\quad$ $\eta = 0.8$

$\rho_a = 600\ kg.m^{-3}$ $\quad$ $D_c = 0.5\ m$

Hence:

$D_i = 0.5/2 = 0.25$ m $\quad$ $r_c = D_c/2 = 0.25$ m

$R_i = D_i/2 = 0.125$ m $\quad$ $L_e = D_c/4 = 0.125$ m

$H = 4D_c = 2$ m $\quad$ $H_e = D_c/2 = 0.25$ m

Cross-section of the inlet:

$$A_e = 0.25 \times 0.125 = 0.03125 \text{ m}^2$$

$$V_e = 0.5/0.03125 = 16 \text{ m.s}^{-1}$$

We can verify that:

$$15 < 16 < 20$$

Cross-section of the chimney:

$$A_i = \pi \times 0.125^2 = 0.0491 \text{ m}^2$$

$$V_i = 0.5/0.0491 = 10.18 \text{ m.s}^{-1}$$

Let us calculate U_i/V_i (see section 4.1.5):

$$Fr = \frac{10.18}{(2 \times 9.81 \times 0.125)^{1/2}} = 6.5$$

$$B = \sqrt{0.8 \times 0.03}\left[\frac{6.5 \times 1.5}{600}\right]^{1/2} \times 0.5^{5/8}$$

$$B = 0.0128$$

$$\frac{U_i}{V_i} = \frac{1}{\dfrac{1.3 \times 0.03125}{\pi \times 0.125 \times 0.25} + \dfrac{2}{0.25}\left[\dfrac{0.01}{2} + 0.0128 \times 0.25\right]}$$

$$\frac{U_i}{V_i} = 2.086$$

$$\varepsilon_S = 0.5\left[\frac{1}{\left[1-\frac{2.086\times 2}{0.125}(0.005+0.0128\times 0.25)\right]^2}-1\right]$$

$$\varepsilon_S = 0.4478$$

$$\varepsilon_R = 1+\frac{2/1.587}{2.086^{0.666}} = 1.772$$

$$\Delta P_G = \frac{1.5}{2}(0.4478+1.772)2.086^2\left[\frac{0.5}{0.0491}\right]^2$$

$$\Delta P_G = 751\,\text{Pa}$$

4.1.8. *Pressure drop in hydrocyclones (approximate calculation)*

Svarowski and Bavishi [SVA 77] proffer the following relation:

$$\Delta P = k\frac{Q_L^{3.26}}{D_c^{4.7}} \qquad \text{(Pascals)}$$

Q_L: feed flowrate ($m^3.s^{-1}$)

D_c: diameter of the cyclone (m)

The coefficient k is expressed by:

$$k = 10^8(4.76 - 0.451\,\text{LnC})$$

C: volumetric fraction of solid in m^3 per m^3 of dispersion

This expression is valid for:

$$0.003 < C < 0.15$$

Remember that the relation between C and the mass loading ratio μ (mass of solid per unit mass of the fluid) is written:

$$C \# \mu \frac{\rho_F}{\rho_S}$$

ρ_F: density of the fluid (kg.m^{-3})

ρ_S: density of the solid (kg.m^{-3})

EXAMPLE 4.2.–

$\rho_L = 1{,}000$ kg.m^{-3} $\mu = 0.03$

$\rho_S = 2{,}000$ kg.m^{-3} $Q_L = 0.005$ m^3.s^{-1}

$D_c = 0.15$ m

$H_e = D/2 = 0.075$ m

$L_e = Dc/4 = 0.0375$ m

$A_e = 0.075 \times 0.0375 = 0.00281$ m^2

$V_e = 0.005/0.00281 = 1.78$ m.s^{-1}

We can verify that:

$$1.5 < 1.78 < 3$$

$$C = 0.03 \times \frac{1{,}000}{2{,}000} = 0.015$$

$$k = 10^8 (4.76 - 0.451 \text{Ln} 0.015) = 2.86 \times 10^8$$

$$\Delta P = 2.86 \times 10^8 \times \frac{0.005^{3.26}}{0.15^{4.7}} = 67{,}231 \text{ Pa}$$

$$\Delta P = 0.67 \text{ bar}$$

NOTE.–

Hydrocyclones can be vulnerable to erosion. It is possible to coat their insides with rubber, polyurethane or corundum cement.

4.1.9. *Cunningham correction*

When the particle diameter is around a micron, the gas can no longer be considered a continuous fluid.

In this case, the mean velocity of the gaseous molecules is:

$$\overline{u} = \sqrt{\frac{8RT}{\pi M}}$$

M: molecular masses of the gas ($kg.kmol^{-1}$)

R: ideal gas constant (8314 $J.kmol^{-1}.K^{-1}$)

T: temperature of the gas (K)

$\overline{u}$ ($m.s^{-1}$)

The mean free path of the gaseous molecules is:

$$\lambda = \frac{\mu_G}{0.5\rho_G\overline{u}}$$

λ: mean free path (m)

μ_G: dynamic viscosity of the gas (Pa.s)

ρ_G: density of the gas ($kg.m^{-3}$)

Cunningham's coefficient is expressed by the following formula, deduced from Davies' semi-empirical expression:

$$C = 1 + \frac{2.52\lambda}{d_p}$$

d_p: diameter of the particles (m)

4.1.10. *Motion of dust particles*

The braking force of the fluid in the laminar regime is:

$$F = 3\pi d_p V_r \mu / C$$

C: Cunningham's corrective coefficient (see section 4.1.9)

V_r: relative velocity of the dust particles with respect to the fluid

If m_p is the mass of a particle, we can write:

$$F = \frac{m_p 3\pi\mu d_p V_r}{C\left(\rho_p - \rho_F\right)\frac{\pi}{6} d_p^3} m_p K V_r$$

where:

$$K = \frac{18\mu}{\left(\rho_p - \rho_F\right) d_p^2 C}$$

Let us project this relation on the three axes in cylindrical coordinates:

(1) $r'' - r\theta'^2 = -Kr'$

(2) $r\theta'' + 2r'\theta' = -Kr(\theta - \phi)$ ($2r'\theta'$ is the Coriolis acceleration)

(3) $z'' = -K(z' - w) - g$

θ: angle of rotation of the dust particles

ϕ: angle of rotation of the fluid

To solve this system, we shall hypothesize that:

$$\theta = \phi = \text{const.} \qquad z' = w$$

Under the terms of this hypothesis, equation [4.2] is reduced to an identity. The condition $z' = w$ is tantamount to discounting the influence of gravity in equation [4.3].

The equilibrium of the forces is written:

$$r'' + Kr' = \frac{V^2}{r}$$

V: orthoradial velocity common to the fluid and the dust particles, which is constant and equal to the velocity of the gas as it enters the cyclone.

In the annular space between r and r + dr, we can consider that the right-hand side of the above equation has a constant value.

This 2nd order linear equation can be easily integrated with the following initial conditions at the inlet to a slice:

$$r = r_1 \qquad r' = r'_1$$

At the outlet from the slice, we obtain the following, by integration, after a time period Δt:

$$r_2 = r_1 + \frac{V^2 \Delta t}{K\bar{r}} + \left[\frac{V^2}{K^2\bar{r}} - \frac{r_1'}{K}\right] e^{-K\Delta t} - \frac{V^2}{K^2\bar{r}} + \frac{r_1'}{K}$$

$$r_2' = \frac{V^2}{K\bar{r}} - \left[\frac{V^2}{K\bar{r}} - r_1\right] e^{-K\Delta t}$$

Thus, it is easy to compute the radial distance of a particle after a time t. At the inlet: $r_1 = r_e$ and $r'_1 = 0$.

In practice, within each slice, we adopt the following value for $\bar{r}$:

$$\bar{r} = \frac{r_1 + r_2}{2}$$

This last condition requires that we perform an iterative calculation in each slice. In the first iteration, we adopt:

$$\bar{r} = r_1$$

Thus, the numerical calculation links the radial distance r of a particle to the time elapsed between its entry and the imposed radial distance r_e. The time step Δt will be taken as equal to:

$$\Delta t = t_{max} / 40$$

We shall now determine t_{max} as regards the length of stay of the fluid filaments.

4.1.11. *Residence time of fluid filaments*

In Figure 4.2, we can see that the fluid flow fluid is directed downwards in the annular space; then the filaments reverse their direction and the flow is directed toward the top of the chimney. The residence time of the filaments varies from zero, for an incoming filament on contact with the chimney (whose radius is r_i), to t_{max} for an incoming filament on contact with the external wall of the cyclone (with radius r_c).

We shall accept that the residence time t_s varies in a linear fashion with the inlet radius r_e of the filament in question.

$$t_s = t_{max} \times \frac{r_e - r_i}{r_c - r_i}$$

However, it may be that we wish to stick very closely to the experimental results. In this case, it would be easy to accept a new matching law between r_e and t_s, e.g. in the form:

$$t_s = t_{max} \left(\frac{r_e - r_i}{r_c - r_i} \right)^n$$

The time t_{max} is the quotient of the total height h of the cyclone by the vertical component of the gas mean velocity in the annular space:

$$t_{max} = \frac{h\pi\left(r_c^2 - r_i^2\right)}{Q_G}$$

Q_G: gaseous flowrate ($m^3.s^{-1}$)

Thus, the residence time t_s for an inlet with radius r_e is:

$$t_s(r_e) = \frac{h\pi}{Q_G}(r_c + r_i)(r_e - r_i)$$

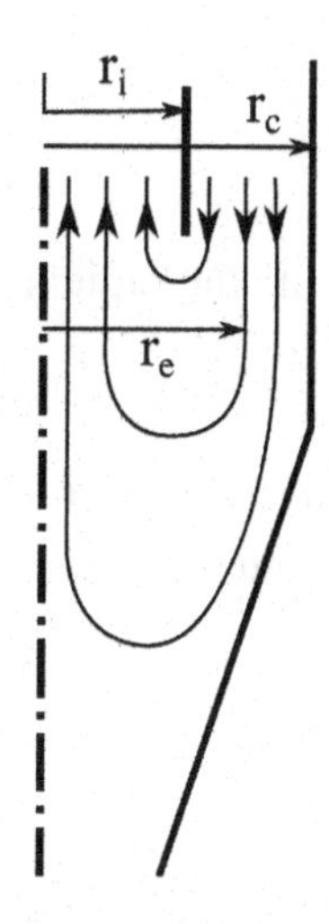

Figure 4.2. *Velocity of the gas in a meridian plane*

4.1.12. *Capture yield for a given particle size*

Taking account of the fact that the inlet into the cyclone is rectangular in cross-section, the capture yield of the cyclone for a particle size d_p is:

$$\eta(d_p) = \frac{r_c - r_e^*}{r_c - r_i}$$

Here, r_e^* is the radial distance at which a particle must enter so that, at the end of the corresponding length of stay r_e^*, the particle just reaches the external wall of the cyclone.

For $r_e^* \leq r_e < r_c$, all the particles are captured.

For $r_i < r_e < r_e^*$, no particle is captured.

To find r_e^*, we need to proceed by successive iterations, integrating the radial motion equation for increasing values of r from $r^{(i)}$. We repeat the operation until the final value r_f of r is equal to r_c with a given degree of accuracy. For this purpose, we proceed as follows:

– we choose the following initial value:

$$r_e^{(1)} = \frac{1}{2}(r_i + r_c)$$

– we then numerically integrate the radial motion equation until:

$$r = r_c$$

This value is reached after a time t_f which is different to the residence time t_s (r_e). We then correct $r_e^{(1)}$ and set:

$$r_e^{(2)} = \frac{r_e^{(1)}}{3}\left(2 + \frac{t_f}{t_s}\right)$$

– we recalculate t_s and integrate again, and so on.

NOTE.–

This method of calculating the yield gives, for the function $\eta(d_p)$, a representative curve which is sigmoidal in shape. However, this S-shaped form is that which is yielded by experience. The results are of the same order of magnitude as those given by the simplified calculation, which we shall examine below. Note, however, that the numerical integration method can be envisaged, not only for gas cyclones but also for hydrocyclones, because it is founded on theoretical considerations.

4.1.13. *Simplified calculation of the yield of a gas cyclone*

We shall use the results found by Rosin *et al.* [ROS 32] adapted by the author of the article on cyclones in the *Chemical Engineers' Handbook* by Perry and Chilton [PER 73].

We begin by calculating the cutoff diameter, meaning the diameter of the particle recovered at 50%:

$$d_{p50} = \left[\frac{9\mu_G L_e}{2\pi N_S V_e (\rho_p - \rho_G)}\right]^{1/2}$$

μ_G: viscosity of the gas (Pa.s)

N_S: number of spires corresponding to the helical motion of the gas

V_e: velocity in the air inlet ($m.s^{-1}$)

L_e: width of the air inlet (m)

ρ_p and ρ_G: densities of the particles and of the gas ($kg.m^{-3}$)

In normal conditions: $N_S = 4$

To express the capture yield $\eta(d_p)$, we can refer again to the *Chemical Engineers' Handbook* [PER 73], which contains an empirical curve that can be expressed analytically, setting $x = d_p/d_{p50}$:

$$0 < x < 1$$

$$\eta(d_p) = 1 - \exp\left[-0.7149 x^{1.6984}\right]$$

$$1 < x < 5$$

$$\eta(d_p) = \frac{1.14285x - 0.5714}{x + 0.14285}$$

$$x > 5$$

$$\eta(d_p) = 1$$

EXAMPLE 4.3.–

$\mu_G = 20\times10^{-6}$ Pa.s $\rho_p = 2{,}000$ kg.m^{-3}

$L_e = 0.125$ m $P_G = 1.5$ kg.m^{-3}

$V_e = 16 \text{ m.s}^{-1}$

$$d_{p50} = \left[\frac{9\times20\times10^{-6}\times0.125}{2\pi\times4\times16\times(2000-1.5)}\right]^{1/2}$$

$$d_{p50} = 5.3\times10^{-6}\text{ m}$$

d_p (µm)	x	η
1	0.19	0.04
3	0.56	0.24
5.3	1	0.49/0.50
10	1.89	0.78
20	3.77	0.95

Table 4.1. *Capture yield of a cyclone*

4.1.14. *Simplified calculation of the yield of a hydrocyclone*

First of all, we calculate the cutoff diameter, drawing inspiration from the work of Trawinski [TRA 69] and Dahlstrom [DAH 54]:

$$d_{p50} = 22.8\times10^{-3} D_c \left[\frac{\mu_A D_c}{Q_A \Delta\rho}\right]^{0.5} \qquad (\text{m})$$

μ_A: viscosity of the feed (Pa.s)

Q_A: feed flowrate (m^3.s^{-1})

Δ_ρ: difference of the densities of the dispersed- and contiguous phases (kg.m^{-3})

To obtain the capture yield, we can, as an initial approach, use the expression advanced by Bradley:

$$\eta(d_p) = 1 - \exp\left[-(x - 0.115)^3\right]$$

where $x = d_p / d_{p50}$

EXAMPLE 4.4.–

$Q_A = 0.05\ m^3.s^{-1}$ $\rho_L = 1000\ kg.m^{-3}$

$\mu_A = 5\times10^{-3}\ Pa.s$ $\rho_S = 2000\ kg.m^{-3}$

$D_c = 0.15\ m$

$$d_{p50} = 22.8\times10^{-3}\times0.15\left[\frac{5\times10^{-3}\times0.15}{0.005(2000-1000)}\right]^{0.5}$$

$$d_{p50} = 42\times10^{-6}\ m$$

On the basis of Bradley's formula [BRA 65], we can construct Table 4.2.

d_p (μm)	x	H
10	0.24	0.002
30	0.71	0.194
42	1	0.50
100	2.38	1

Table 4.2. *Cyclone capture yield simplified calculation*

NOTE.–

In the general case of a widespread PSD, we can construct the following table:

Column 1: mean diameter $\overline{d_{pi}}$ of each size class of the solid material to be recovered (geometric mean of the two extreme diameters).

Column 2: mass fraction of each class in total.

Column 3: ratio of $\overline{d_{pi}}/d_{p50}$ (simplified method) or value of $r_e^*(d_p)$ (numerical integration method).

Column 4: capture yield of each class.

Column 5: recovered fractions (column 2 multiplied by column 4).

By totaling the figures in column 5, we obtain the overall yield of the cyclone.

4.1.15. *Wet aerosols*

A wet aerosol is a suspension of liquid droplets in a gas. In principle, it is possible to design a cyclone to treat these aerosols. The conical part of the classic cyclone (i.e. for dry aerosols), whose height is equal to $2D_c$, will be replaced by a cylinder whose height is D_c (D_c is the diameter of the cyclone).

Indeed, the conical form is necessary for the evacuation of the dry solid particles at the bottom. On the other hand, to evacuate the liquid having accumulated on the wall of the cyclone, we merely need a hole in the flat bottom of the cylindrical part. This is advantageous because a flat-bottomed cylinder is less costly to manufacture than a cone.

Despite all of this, such flat-bottomed cyclones are infrequently manufactured, because in general it is not easy to obtain an exact value of the PSD of the droplets, and to rid a gas of the liquid vesicles which it contains in suspension, it is preferable to filter the gas through a droplet-separating mat.

4.1.16. *Conclusions*

Although it is not always apparent from the mathematical expressions used, it must be remembered that:

– the pressure drop decreases slightly when the load in terms of particles increases, because the particles lubricate the friction of the fluid against the walls;

– the capture yield increases slightly with the load of particles, because re-entrainment decreases. Indeed, as before, friction with the wall is reduced.

These devices are simple, and therefore cheap. For this reason, they are very widely used:

– gas cyclones are used in installations for drying divided solids, and also for the recovery of the catalyst in fluidized-bed reactors and also at the peak of aerated silos, etc;

– hydrocyclones are used in wet ore-milling circuits (e.g. for sand removal) and also to clarify certain food liquids, etc.

4.2. The disk decanter

4.2.1. *Description*

This equipment is available in several very distinct versions. Here, we shall examine a centrifugal decanter designed to separate the components of a liquid–solid dispersion where the separated solid accumulates in the bowl. This is the solids-retention device. This device is a clarifier (it clarifies the liquid).

The suspension enters at the top of the device and, after an axial journey, is conveyed to the edge of the disks (also known as plates) by a conical system of radial bars which are not shown in Figure 4.3. In almost all cases, the solid is denser than the liquid and, under the influence of the centrifugal force, it is deposited on the underside of the disks.

The mud thus deposited travels toward the periphery and accumulates on the wall of the bowl. Therefore, the bowl needs to be periodically dismantled and cleaned. The clarified liquid, for its part, travels across the surfaces of the disks and is evacuated by a device which we shall not describe here.

The number of disks can vary from 50 to 150, and their vertical spacing, which is generally around 2 mm, may be up to 10 mm for a product that does not flow easily.

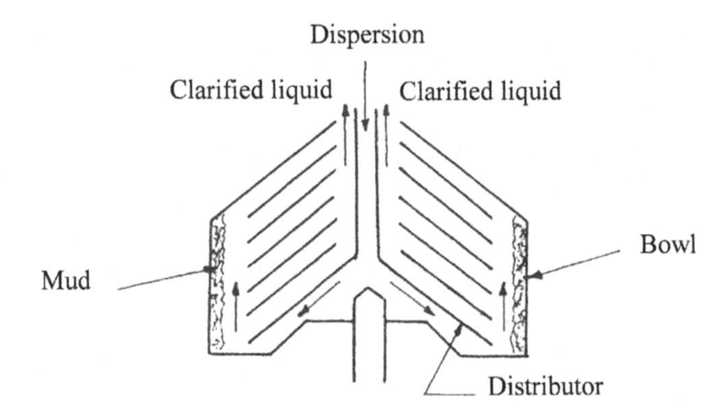

Figure 4.3. *Solids-retention decanter*

NOTE.–

The dispersion decants, progressing toward the axis. If precautions are not taken, the Coriolis acceleration resulting from the combination of the rotation and that centripetal displacement lends the suspension a spiral motion with a peripheral (orthoradial) velocity of around 1 $m.s^{-1}$. Such a velocity would cause the re-entrainment of the decanted mud, which would significantly decrease the separation yield of the device. For this reason, the disks have radial bars which act as thickness wedges when the disks are stacked. These bars prevent orthoradial motion of the suspension, and the re-entrainment is practically zero because the velocity of the suspension in relation to the disks is reduced to its radial component, which is a few $cm.s^{-1}$. Such a device is said to operate in meridian flow. The number of sectors delimited by the bars may, depending on the case, be between 6 and 12.

4.2.2. *Velocity profile*

Between two disks, mud accumulates on the upper disk and reduces the space available for the displacement of the clear liquid.

Hereafter, we shall accept that the above velocity profile is also that of the particles.

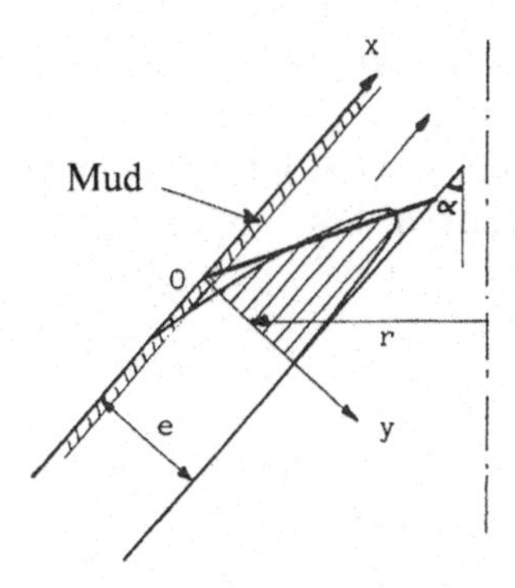

Figure 4.4. *Velocity profile*

Owing to the braking force exerted by the sediment flowing in the opposite direction to the clear liquid, the latter's velocity profile is deformed, and is different from the usual parabolic profile for the laminar flow of a liquid between two plates. As Figure 4.4 shows, we can assimilate that profile to a straight line, so the velocity u, for our purposes, will be proportional to the y coordinate:

$$u = C\,y$$

This velocity profile must be able to carry the flowrate Q:

$$Q = 2\pi r \int_0^e u\,dy = 2\pi r \frac{C}{2} e^2$$

Thus:

$$C = \frac{Q}{\pi r e^2}$$

$$u = \frac{Q}{\pi r e^2} y$$

4.2.3. *Theoretical separation yield*

In the direction of the Oy axis, the displacement of a particle over the time dεr is, according to Stokes' law:

$$dy = \cos\alpha\ \omega^2\ r\tau_o d\tau \qquad [4.7]$$

The factor $\cos\alpha\ \omega^2$ is the component of the centrifugal acceleration normal to the surface of the plates.

The factor τr_o, which has the dimensions of a period of time, characterizes the suspension:

$$\tau_o = \frac{d_p{}^2 \Delta\rho}{18\mu} \qquad [4.8]$$

d_p: diameter of the particles (m)

$\Delta\rho$: difference between the densities of the solid and the liquid ($kg.m^{-3}$)

μ: viscosity of the liquid phase (Pa.s)

Also, we can see in Figure 4.4 that:

$$dx = -\frac{dr}{\sin\alpha} = ud\tau = \frac{Q}{\pi re^2} yd\tau \qquad [4.9]$$

Let us eliminate dεr between equations [4.7] and [4.9]:

$$r^2dr = \frac{2Qtg\alpha}{2\pi e^2\omega^2\tau_0} y\ d\ y$$

If r_e and r_i are the radii corresponding to the exterior and the interior of the stack of plates, let us look for the ordinate value y* at which a particle needs to be at the radial distance r_e so that y reaches zero when the particle has arrived at the radial distance r_i. For this purpose, we merely need to integrate the above equation:

$$\frac{r_i^3 - r_e^3}{3} = \frac{Qtg\ \alpha y^{*2}}{2\pi re^2\omega^2\tau_0}$$

and, when we replace er_0 with its value:

$$d_p^2 = \frac{27\mu tg\alpha Q}{\left(r_e^3 - r_i^3\right)\pi\omega^2\Delta\rho}\left[\frac{y^*}{e}\right]^2$$

If a particle with diameter d_p, when it penetrates into the interval whose width is e, has a y coordinate $y < y^*$, then it will surely be separated. Put differently, the fraction of the feed which will be separated will be:

$$\eta = \frac{y^*}{e}$$

The variable η is the separation yield.

We can see that the yield η is proportional to the diameter d_p of the particles but, of course, it cannot surpass the value of 1. For this maximum value there is a corresponding diameter, which is that of the "limiting particle" which is entirely separated.

EXAMPLE 4.5.–

$r_e = 0.1$ m	$Q = 5.1\times10^{-6}$ m^3.s^{-1}	$a = 35°$ sexagesimal
$r_i = 0.03$ m	$\Delta\rho = 180$ kg.m^{-3}	tg $\alpha = 0.7$
$e = 0.001$ m	$\mu = 10^{-3}$ Pa.s	$\omega = 418$ rad.s^{-1}

Let us try to find the diameter of the limiting particle:

$$d_p^* = \left[\frac{27\times10^{-3}\times0.7\times5.1\times10^{-6}}{\left(0.1^3 - 0.03^3\right)\pi\times418^2\times180}\right]^{1/2}$$

$$d_p^* = 0.99\times10^{-6}\text{ m}$$

NOTE.–

We have determined $\eta = f(d_p)$ on the basis of a velocity profile but, conversely, it is possible to calculate a velocity profile on the basis of a function $d_p = g(\eta)$ found by experimentation.

We could, for the dispersion, have adopted a velocity profile between two disks which is symmetrical in relation to the two disks, and approaches the parabolic profile of the laminar flow, but the calculation would have shown that such a profile leads to a simply absurd yield function $\eta(d_p)$.

4.2.4. *Evolution of the theoretical yield as a function of the flowrate*

We have seen that:

$$\eta = \frac{y^*}{e} = \frac{K}{Q^{0.5}} d_p$$

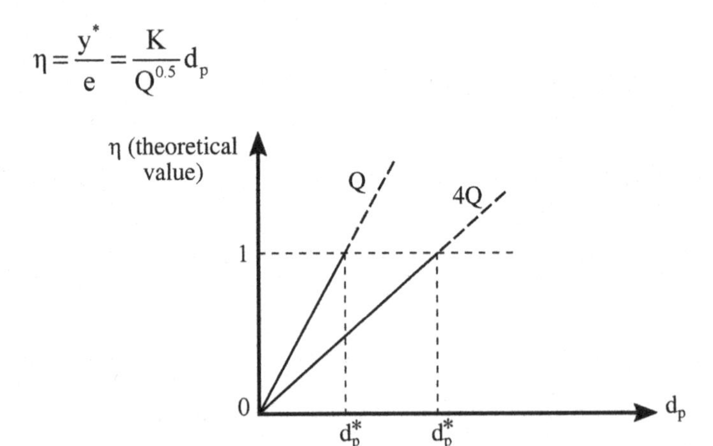

Figure 4.5. *Variation of η as a function of the flowrate*

If we multiply the flowrate by 4, the limiting particle size for which $\eta = 1$ is multiplied by 2.

4.2.5. *True separation yield*

At the point where the dispersion enters the space between the disks, relative to a sector, vorticial motion is created, due to the Coriolis acceleration. This phenomenon tends to render the medium homogeneous, and hampers sedimentation on the edge of the disks. This effect extends over the radial distance Δx above the lower disk, but has no influence on the upper disk where the mud is flowing because, as we have seen, in the vicinity of that disk, the radial velocity of the suspension tends toward zero, and so too does the Coriolis acceleration. We shall suppose that the neutralized radial distance decreases as a linear function of Δx, falling to zero when y decreases from e to 0.

Begin by considering the case of a particle whose diameter is the limiting diameter, for which the theoretical yield is equal to 1. In the presence of disturbances at the inlet, such a particle, entering at y_2, is simply decanted after passing through the space between the disks. Owing to the existence of the inlet eddy, its separation yield, which was equal to 1, becomes less than 1 and equal to y_2/e.

A particle much larger than the one discussed above will decant in any case, so the inlet effect will have no influence.

A particle smaller than the limiting particle should return to y_1 before decanting on exiting the inter-disk space. In Figure 4.6, we can see that, at this value of y, the influence of the inlet disturbance is maximal in the vicinity of the limiting particle. The shape of the actual yield is shown by Figure 4.7 (curve (1)).

Note that if the liquid in the dispersion is significantly more viscous than water, the rotational motion of the eddy will be slowed, and the yield will improve, as shown by the dashed curve (2).

The above points account for the experimental results obtained by Brunner and Molerus [BRU 80].

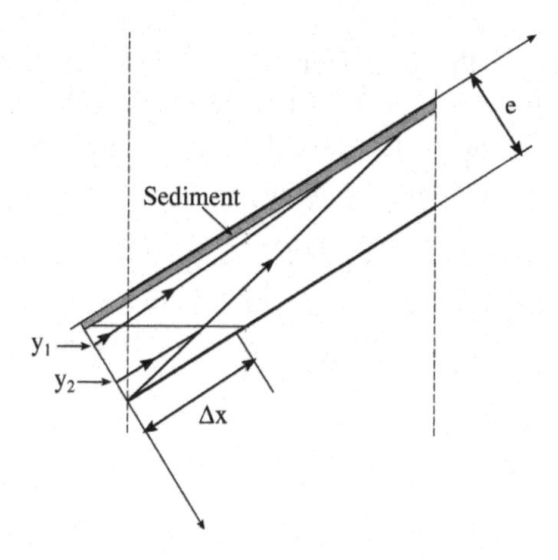

Figure 4.6. *Influence of the inlet eddy*

4.2.6. *Engorgement*

When running the calculations for falling-film evaporators, we are able to evaluate the thickness of the film. That thickness is a function of the viscosity of the liquid at hand and the acceleration due to gravity g. The film of decanted mud, flowing over the lower surface of the upper disk in the inter-disk space, flows in the same way, and its thickness can be estimated by replacing g by $\omega^2 r_e \sin \alpha$ and adopting a value of around 1 Pa.s for the viscosity of the mud deposited.

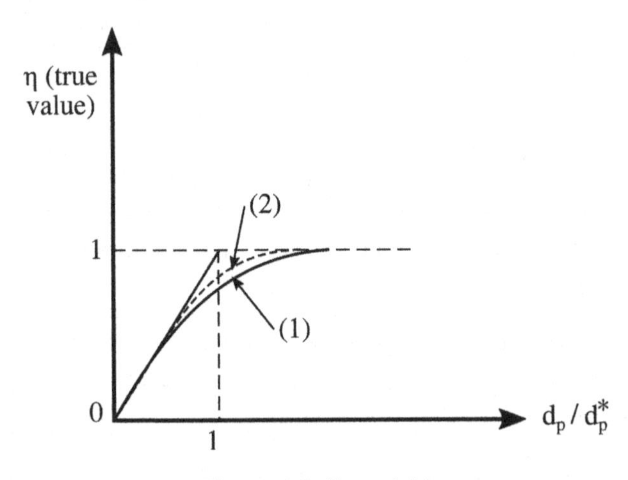

Figure 4.7. *True yield*

As the thickness of the film increases with the flowrate of mud to the power of 1/3, may for a certain feed flowrate, that thickness reach the value of the inter-disk space, in which case the dispersion can no longer flow and we see the phenomenon of engorgement.

Thus, for highly viscous gum arabic, the thickness of the film is such that if we want the suspension to be able to flow between the film of gum and the lower disk, we must leave an inter-disk space of 10 mm.

4.2.7. *Use of disk centrifuge decanters*

These devices are appropriate when the size of the particles is between 0.5 and 30 µm. They work well for particles larger than 30 µm, but there are more economical methods for this.

Pressurized filtration, often with an adjuvant, in the range of sizes from 1 to 30 μm can compete with centrifugal decantation. However, disk devices must be used in the following cases:

– the particles are deformable (as is the case with micro-organisms) and, over the course of a filtration operation, form not a cake but a true gel, resistant to the passage of the filtrate;

– the liquid phase is relatively viscous; such is the case of food oils or mineral oils;

– the solid is the noble product. If the particles are in the form of a fiber, a filtration cake will have a porosity greater than 0.95 and filtration will be possible. On the other hand, if the particles are spheroidal, the porosity of the cake may drop to 0.4, and if, in addition, the size of the particles approaches 1 μm, a filtration adjuvant will be necessary, which pollutes the solid we need to recover. Such is the case of dyes and certain pharmaceutical products.

Solids-retention devices can only be used for dispersions containing less than 1% solid per volume, so that the bowl does not require cleaning excessively often.

There are automatic mud-discharge devices using peripheral buses. Such devices can comfortably handle dispersions containing up to 25% of solids.

An interesting use of disk decanters is the fractionation of emulsions, i.e. of dispersions of droplets of an insoluble liquid in another liquid. In this case, the emulsion is not introduced between the disks from their periphery. In each disk, three or four holes are made, oriented 120° or 90° apart, superposed from one disk to the next to form three or four distribution columns. Naturally, the lighter of the liquids runs toward the axis, leaving the heavier liquid behind, whilst the heavier liquid flows toward the periphery, separating from the lighter component.

If the holes are near the periphery, we say we are dealing with a purifier, because the disks essentially serve to purify the lightweight liquid. On the other hand, if the holes are near to the axis, then we are dealing with a concentrator, which increases the concentration of the heavier liquid. For the calculation:

– in the case of a purifier, we take the value of r_e, which corresponds to the position of the distribution columns;

– in the case of a concentrator, we take the value of r_i which is defined by the position of the distribution columns.

In closing, note that a large disk decanter can treat up to $40 m^3.h^{-1}$ of dispersion.

4.3. Tubular decanter

4.3.1. *Description*

The bowl of the tubular decanter is a simple cylinder with a vertical axis. The separated solid accumulates along the lateral wall, from which it must be extracted after interruption of the feed.

The dispersion enters at one end of the bowl and the clarified liquid flows out at the other end, passing through a circular spillway called a diaphragm.

This device has a low capacity because the maximum dispersion flowrate it is able to treat is 200 liters/h, whereas a disk decanter can accept up to $40 m^3.h^{-1}$. On the other hand, the tubular decanter can rotate much faster (15,000 to 20,000 $rev.mn^{-1}$) than disk decanters (3,000 to 4,000 $rev.mn^{-1}$).

4.3.2. *The separation yield*

After entering the bowl, the suspension has a uniform axial velocity after having covered the distance ΔL_e. Here, we discount the Coriolis force. At the outlet, the clarified liquid crosses the diaphragm whose radius is r_D after approaching the axis along the length ΔL_S.

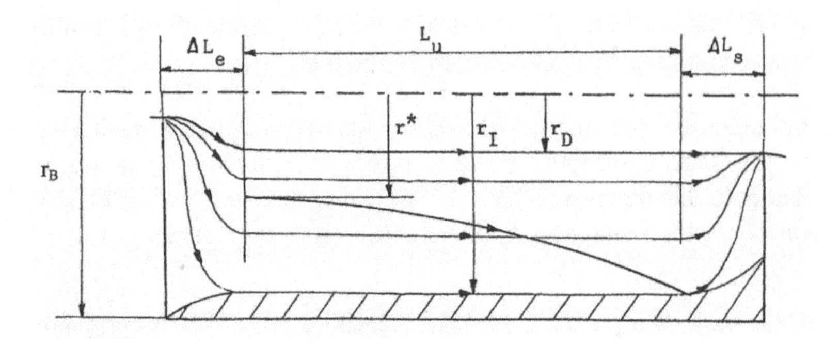

Figure 4.8. *Theoretical trajectory of a particle in a tubular decanter*

Thus, centrifugal decantation takes place only in the so-called effective zone – i.e. over the usable length L_u. The area available for the passage of the suspension is a circular corona measuring between the radius r_D of the diaphragm and the radius r_I of the interface separating the suspension and the deposited solid.

A particle which penetrates the effective zone at the radial distance r* will be separated if it has reached the surface of the solid deposit whose radius is r_I after having covered the axial distance L_u.

Stokes' law, applied to a particle subject to centrifugal force, can be used to express its limiting velocity:

$$\frac{dr}{d\tau} = \frac{\omega^2 r \Delta\rho d_p^2}{18\mu}$$

ω: angular speed of rotation of the bowl ($rad.s^{-1}$)

r: radial distance of the particle (m)

Δρ: difference between the densities of the particle and of the clear liquid ($kg.m^{-3}$)

d_p: diameter of the particle (m)

μ: viscosity of the clear liquid (Pa.s)

We set:

$$\tau_o = d_p^2 \Delta\rho / (18\mu)$$

Stokes' law is then written:

$$dr = \omega^2 r \tau_o d\tau \qquad [4.10]$$

Additionally, the axial velocity of the suspension is:

$$V_A = \frac{Q}{\pi\left(r_I^2 - r_D^2\right)}$$

The elementary axial displacement of the particle is:

$$dL = V_A d\tau \qquad [4.11]$$

By eliminating $d\tau$ between equations [4.10] and [4.11] and integrating from r_o to r_I, we obtain:

$$\mathrm{Ln}\frac{r_I}{r^*} = \frac{\omega^2 \tau_o L_u}{V_A} \qquad [4.12]$$

A particle of the same diameter d_p which penetrates into the effective zone at a radial distance radial less than r* will not be separated. On the other hand, if the radial distance from the inlet is greater than r*, the particle will be separated. This gives us the separation yield:

$$\eta = \frac{r_I^2 - r^{*2}}{r_I^2 - r_D^2}$$

Those particles which are separated completely, then, will be such that $r^* = r_D$, meaning that, in view of equation [4.12]:

$$\tau_o = \frac{V_A}{L_u \omega^2} \mathrm{Ln}\frac{r_I}{r_D}$$

The parameter characteristic of the solid phase is then:

$$d_p^2 \Delta\rho = \frac{18\mu V_A}{L_u \omega^2} \mathrm{Ln}\frac{r_I}{r_D}$$

For a suspension which has a small load, we can assimilate r_I to the radius of the bowl r_B.

EXAMPLE 4.6.–

$r_B = 0.04$ m $\quad Q = 183$ L/h $= 5\times10^{-4}$ m^3.s^{-1} $\quad \omega = 1500$ rad.s^{-1}

$r_D = 0.03$ m $\quad \mu = 10^{-3}$ Pa.s $\quad$ Lu $= 0.3$ m

$$V_A = \frac{5\times10^{-4}}{\pi\left(0.04^2 - 0.03^2\right)} = 0.227\ \text{m.s}^{-1}$$

$$d_p^2\Delta\rho = \frac{18\times10^{-3}\times0.227}{0.3\times1500^2}\text{Ln}\frac{0.04}{0.03} = 1.74\times10^{-9}\ \text{kg.m}^{-1}$$

and, if $\Delta\rho = 50$ kg.m^{-3}

$$d_p = 5.9\ \mu\text{m}$$

4.4. Screw mud separator

4.4.1. *Operational principles (Figure 4.9)*

The above example shows that the performances of the tubular decanter are not exceptional, not only from the point of view of the flowrate processed, but also in terms of the size of the particles separated. Indeed, it develops a limited centrifugal acceleration because its bowl has a small diameter. By increasing the size of the device, we must therefore improve its performances.

The industrial version is on a horizontal axis and its bowl rotates at a variable velocity depending on the machine between 1,200 and 3,000 rev.mn^{-1}. The upper bound of the diameter of the bowl is of the order of 2 m. This machine is called a mud separator, because it is widely used to clarify aqueous effluents by stripping them of the mud they contain.

The volume of the thickened mud must be no greater than 40% of the volume of the suspension undergoing treatment. This limit is significantly higher than for the disk decanter or the tubular decanter. Note, though, that in reality there are three ways of assessing the characteristics of a suspension relative to centrifugal decantation:

1) the dryness is the fraction of dry material (in terms of mass) remaining after complete desiccation in a drying room, at a temperature of around 80°C;

2) the volumetric fraction of thickened mud, which is measured by centrifuging a graduated flask filled with suspension and finding the relation

between the volume of the deposited residue and the volume of the suspension present in the graduated flask;

3) the volumetric fraction of solid in the suspension. This value can only be found by centrifugal- or pressurized filtration. It is the only parameter which can characterize a suspension rationally. However, whilst it is less easy to measure, it is more commonly measured.

The mud separator has a screw which is coaxial to the bowl, rotating at a slightly different velocity to that of the bowl. The difference between these two velocities is the relative velocity of the screw, which may range between 1 and 25 rev.mn^{-1}. That screw entrains the thickened mud toward the opposite end from the outlet for the clarified liquid, as is shown by Figure 4.9.

The clarified liquid exits through a diaphragm, as it does in the case of the tubular decanter. It is customary to measure the hydraulic flowrate, which is the maximum flowrate of pure water that the machine can accommodate without that water exiting through the mud outlet. This flowrate is much higher than the flowrate compatible with a satisfactory separation of the solid in suspension.

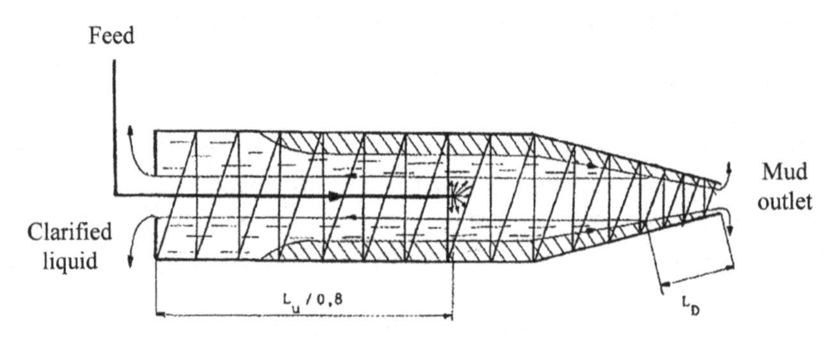

Figure 4.9. *Principle of the mud remover*

Around the mud outlet, the bowl takes the form of a truncated cone, which forces the mud to move toward the axis of the bowl and emerge on the free surface of the suspension. This zone has the length L_D if measured along a generatrix of the cone trunk, and it is here that the drainage and final thickening of the mud takes place. If the feed has a high proportion of solid, the thickness of mud deposited on the wall of the bowl will be significant, and will increase when there is high resistant torque for the rotation of the screw relative to the bowl. To limit the mass of mud present in the machine, i.e. to decrease the resistant torque, we need to increase the relative velocity of the screw. However, consequently the mud will stay in the machine for less time, will be less compacted, less well drained and will be wetter as it exits. Hydraulic or electronic devices can automatically modify the relative velocity ω_R as a function of the resistant torque C. It is possible to set the sensitivity of that regulation – i.e. the ratio $d\omega_R/dC$ – in advance.

4.4.2. *Velocity of the clarified liquid between the threads of the screw*

The mud slips over the surface of the threads of the screw, and moves only in the direction of the axis of the bowl. Its flowrate is:

$$Q_B = N_{VR} p\pi\left(r_B^2 - r_I^2\right)$$

N_{VR}: relative speed of rotation of the screw in relation to the bowl ($rev.s^{-1}$)

p: step of the screw (m)

r_B: internal radius of the bowl (m)

r_I: radius of the interface separating the thickened mud from the liquid (m)

This relation can be used to calculate r_I when we know the flowrate Q_B, which is, itself, the product of the feed flowrate by the volumetric fraction of thickened mud (measured with a bladed centrifuge).

The angle α of inclination of the threads on the axis is given by:

$$\tan\alpha = 2\pi r_B / p$$

The slip velocity of the mud in relation to the threads is deduced from this:

$$V_{BF} = \frac{2\pi N_{VR} r_I}{\sin\alpha}$$

Let us now try to find the velocity of the clarified liquid in relation to the threads. The axial flowrate Q_L of that liquid is:

$$Q_L = (\text{flowrate in relation to the threads}) - (\text{axial translation due to the motion of the screw})$$

That is:

$$Q_L = V_{LF} p \sin\alpha (r_I - r_D) - N_{VR} p\pi (r_I^2 - r_D^2)$$

Thus:

$$V_{LF} = \frac{Q_L + N_{VR} p\pi (r_I^2 - r_D^2)}{p \sin\alpha (r_I - r_D)}$$

In order to determine the velocity profile of the liquid between the threads, we shall use an example.

EXAMPLE 4.7.–

$r_B = 0.25$ m $Q_L = 2.77\times10^{-3}$ m^3.s^{-1}

$r_D = 0.15$ m $Q_B = 0.554\times10^{-3}$ m^3.s^{-1}

$p = 0.15$ m $N_{VR} = 0.1$ rev.s^{-1}

$$\rho_L = 1{,}000 \text{ kg.m}^{-3}$$

$$r_I^2 = 0.25^2 - \frac{0.554\times10^{-3}}{0.1\times\pi\times0.15} = 0.05 \text{ m}^2$$

$$r_I = 0.225\text{m}$$

$$\cos\alpha = \frac{0.15}{2\pi\times0.25} = 0.0955$$

$$\alpha = 90° - 5.45° = 84.55° \qquad \sin\alpha = 0.995$$

The velocity of the liquid in relation to the threads is:

$$V_{LF} = \frac{2.77\times10^{-3} + 0.1\times0.15\times\pi\left(0.225^2 - 0.15^2\right)}{0.15\times0.995\left(0.225 - 0.15\right)}$$

$$V_{LF} = 0.365 \text{ m.s}^{-1}$$

The perimeter wetted by the liquid is:

$$P = p\sin\alpha + 2\left(r_I - r_D\right) = 0.15\times0.995 + 2\left(0.225 - 0.15\right)$$

$$P = 0.3 \text{ m}$$

The cross-section available for the passage of the liquid is:

$$A = p\sin\alpha\left(r_I - r_D\right) = 0.15\times0.995\left(0.225 - 0.15\right)$$

$$A = 0.0112 \text{ m}^2$$

This gives us the equivalent diameter:

$$D_{eq} = 4A/P = 4\times0.0112/0.3 = 0.15 \text{ m}$$

The Reynolds number for the flow of the liquid is deduced, using the assumption that the viscosity of the liquid is 10^{-3} Pa.s and that its density is 1,000 kg.m^{-3}:

$$Re = \frac{V_{LF}D_{eq}\rho_L}{\mu} = \frac{0.365\times0.15\times1{,}000}{10^{-3}} = 54{,}750$$

4.4.3. *Conclusions*

The value found above for the Reynolds number shows that the flow is certainly turbulent. Thus, we can assume a flat velocity profile between the threads.

In addition, the calculation would show that the velocity of the liquid in relation to the mud is:

$$V_{LF} + V_{BF} = 0.365 + 0.142 = 0.5 \text{ m.s}^{-1}$$

Thus, the mud is re-entrained by the liquid, but the re-entrained mass is slight, if not negligible, in comparison to the mass of mud, unlike in the case with disk decanters without containment bars. Indeed, the acceptable content of solid in the feed of a mud pump is much higher than the content allowable for a disk decanter.

4.4.4. *Performances of the mud pump*

These performances are judged on the basis not of a single criterion, but of two – namely:

– the yield in terms of separation of the solid;

– the dryness of the mud obtained.

1) Separation:

Since the velocity profile of the liquid is flat between the threads of the screw, the presence of the screw causes only a slow overall rotation of the liquid in relation to the bowl, and decantation takes place in the same way as if the liquid were moving simply in parallel to the axis, rotating at the same velocity as the bowl. The separation yield can then be evaluated in the same way as for a tubular decanter, provided the usable length is chosen as 80% of the axial distance between the feed introduction point and the outlet diaphragm.

2) Dryness:

The dryness of the thickened mud depends on its aptitude to be compressed under the influence of the centrifugal forcc. Thc *a priori* calculations are complex and are rarely performed.

EXAMPLE 4.8.–

$r_D = 0.15$ m $\quad L = 1$ m $\quad Q = 2.77 \times 10^{-3}$ m^3.s^{-1}

$r_I = 0.22$ m $\quad \omega = 210$ rad.s^{-1} $\quad \mu = 10^{-3}$ Pa.s

$$V_A = \frac{2.77\times10^{-3}}{\pi\left(0.22^2 - 0.15^2\right)} = 0.034 \text{ m.s}^{-1}$$

$$d_p^2\Delta\rho = \frac{18\times10^{-3}\times0.034}{1\times210^2}\text{Ln}\frac{0.22}{0.15}$$

$$d_p^2\Delta\rho = 0.53\times10^{-8} \text{ kg.m}^{-1}$$

If: $\Delta\rho = 50$ kg.m^{-3}

$$d_p = 10.2\ \mu\text{m}$$

APPENDICES

Appendix 1

Numerical Integration: the Fourth-order Runge–Kutta Method

The aim here is to integrate the differential equation:

$$\frac{dx}{d\tau} = F(x, \tau)$$

$$x_{\tau=0} = x_0$$

We set:

$$\tau_{i+1/2} = \tau_i + \frac{\Delta\tau}{2}$$

$$x_{i+1/2}^{(1)} = x_i + \frac{\Delta\tau}{2} F(x_i, \tau_i)$$

$$x_{i+1/2}^{(2)} = x_i + \frac{\Delta\tau}{2} F(x_{i+1/2}^{(1)}, \tau_{i+1/2})$$

$$x_{i+1}^{(1)} = x_i + \Delta\tau\, F(x_{i+1/2}^{(2)}, \tau_{i+1/2})$$

Therefore:

$$x_{i+1} = x_i + \frac{\Delta\tau}{6}\Big[F(x_i, \tau_i) + 2F(x_{i+1/2}^{(1)}, \tau_{i+1/2}) + 2F(x_{i+1/2}^{(2)}, \tau_{i+1/2}) + F(x_{i+1}^{(1)}, \tau_{i+1})\Big]$$

The method can be generalized to apply to a system of n 1st-order differential equations involving n variables x_j (j ranging from 1 to n). The independent variable is x_0.

$$\frac{dx_j}{dx_0} = F_j\left(x_0, x_1, \ldots, x_j, \ldots, x_n\right)$$

Let us set:

$$x_{0,i+1} = x_{0,i} + \Delta x_0 \quad \text{and} \quad x_{0,i+1/2} = x_{0,i} + \frac{\Delta x_0}{2}$$

$$x_{j,i+1/2}^{(1)} = x_{j,i} + \frac{\Delta x_0}{2} F_j\left(x_{0,1}, \ldots, x_{j,i}, \ldots x_{n,i}\right)$$

$$x_{j,i+1/2}^{(2)} = x_{j,i} + \frac{\Delta x_0}{2} F_j\left(x_{0,i+1/2}, \ldots, x_{j,i+1/2}^{(1)}, \ldots, x_{n,i+1/2}^{(1)}\right)$$

$$x_{j,i+1}^{(1)} = x_{j,i} + \Delta x_0 F\left(x_{0,i+1/2}, \ldots, x_{j,i+1/2}^{(2)}, \ldots, x_{n,i+1/2}^{(2)}\right)$$

and finally:

$$x_{j,i+1} = x_{j,i} + \frac{\Delta x_0}{6}\Big[F_j\left(x_{0,i}, \ldots x_{j,i}, \ldots, x_{n,i}\right) + 2F_j\left(x_{0,i+1/2}, \ldots x_{j,i+1/2}^{(1)}, \ldots x_{n,i+1/2}^{(1)}\right)$$
$$+ 2F_j\left(x_{0,i+1/2}, \ldots x_{j,i+1/2}^{(2)}, \ldots x_{n,i+1/2}^{(2)}\right) + F_j\left(x_{0,i+1}, \ldots x_{j,i+1}^{(1)}, \ldots x_{n,i+1}^{(1)}\right)\Big]$$

Appendix 2

The Cgs Electromagnetic System

A2.1. Potential in the international system (SI) and in the CGS electromagnetic system

By definition (if 1 C = 1 Coulomb)

$$1\underset{\mathrm{SI}}{\mathrm{C}} = 3.10^{9}\, q_{\mathrm{CGSEM}} \text{ and } \varepsilon_0 = \frac{1}{36\pi.10^{9}}\,\mathrm{C.m^{-1}.V^{-1}}$$

The potential created in a vacuum by a 1C charge at a distance of 1 m is:

$$\mathrm{Pot}_{\mathrm{SI}} = \frac{1\,\mathrm{C}}{4\pi\varepsilon_0 1\,\mathrm{m}} = 9.10^{9}\ \mathrm{Volt}$$

The potential created by a charge q of 1 CGSEM unit at the distance of 1cm is:

$$\mathrm{Pot}_{\mathrm{CGSEM}} = \frac{q}{1\ \mathrm{cm}} = 1\ \text{volt stat}$$

The ratio between these two potentials must be such that:

$$\frac{\mathrm{Pot}_{\mathrm{SI}}}{\mathrm{Pot}_{\mathrm{EMCGS}}} = \frac{\mathrm{C}}{q} \times \frac{\mathrm{cm}}{\mathrm{m}}$$

Put differently:

$$\frac{9.10^9\,\text{Volt}}{1\,\text{volt stat}} = \frac{C}{q}\frac{cm}{m} = \frac{3.10^9}{100}$$

Thus:

$$1V = \frac{1\,\text{volt stat}}{300}$$

NOTE.–

We can very simply write:

$$\text{Volt} \times C = \text{Joule}$$

$$\text{Volt stat} \times q = \text{erg}$$

Therefore:

$$\frac{1\,\text{Volt}}{1\,\text{volt stat}} = \frac{\text{Joule}}{\text{erg}} \times \frac{q}{C} = 10^7 \times \frac{1}{3.10^9} = \frac{1}{300}$$

A2.2. Other units in SI and CGSEM

1) The conductivity κ is measured in $\Omega^{-1}.m^{-1}$. The conductance is the quotient of the intensity by the potential.

Finally:

$$[\kappa] = \frac{I}{VL} = \frac{C}{V.s.m} \qquad \text{(SI)}$$

However:

$$1\text{ C} = 3.10^9\, q_{CGSEM}$$

$$1\text{ V} = \frac{1}{300}\,\text{volt stat}$$

$$1\text{ m} = 100\text{ cm}$$

Hence:

$$\kappa_{SI} = \frac{3.10^9 \times 300}{100} \kappa_{CGSEM} = 9.10^9 \kappa_{CGSEM}$$

2) Although viscosity is not an electromagnetic property, its correspondence between the CGS system and the SI is given here. Viscosity is measured in Pa.s, that is to say $kg.m^{-1}.s^{-1}$.

However:

$$1 \text{ kg} = 1000 \text{ g}$$

$$1 \text{ m} = 100 \text{ cm}$$

Therefore:

$$1 \text{ Pa.s} = 10 \text{ g. cm}^{-1}.\text{s}^{-1} = 10 \text{ poise} = 10 \text{ barye.s}$$

Appendix 3

Mohs Scale

Nature of the divided solid	*Mohs hardness*
Wax	0.02
Graphite	0.5–1
Talc	1
Diatomaceous earth	1–1.5
Asphalt	1.5
Lead	1.5
Gypsum	2
Human nail	2
Organic crystals	2
Flaked sodium carbonate	2
Sulfur	2
Salt	2
Tin	2
Zinc	2
Anthracite	2.2
Silver	2.5
Borax	2.5
Kaolin	2.5
Litharge (yellow lead)	2.5
Sodium bicarbonate	2.5
Copper (coins)	2.5

Slaked lime	2–3
Aluminum	2–3
Quicklime	2–4
Calcite	3
Bauxite	3
Mica	3
Plastic materials	3
Barite	3.3
Brass	3–4
Limestone	3–4
Dolomite	3.5–4
Siderite	3.5–4
Sphalerite	3.5–4
Chalcopyrite	3.5–4
Fluorite	4
Pyrrhotite	4
Iron	4–5
Zinc oxide	4.5
Glass	4.5–6.5
Apatite	5
Carbon black	5
Asbestos	5
Steel	5–8.5
Chromite	5.5
Magnetite	6
Orthoclase	6
Clinker	6
Iron oxide	6
Feldspar	6
Pumice stone	6
Magnesia (MgO)	5–6.5
Pyrite	6.5
Titanium oxide	6.5
Quartz	7

Sand	7
Zirconia	7
Beryl	7
Topaz	8
Emery	7–9
Garnet	8.2
Sapphire	9
Corrundum	9
Tungsten carbide	9.2
Alumina	9.25
Tantalum carbide	9.3
Titanium carbide	9.4
Silicon carbide	9.4
Boron carbide	9.5
Diamond	10

Thus, materials can be classified according to their hardness:

Soft 1 to 3

Fairly soft 4 to 6

Hard 7 to 10

Appendix 4

Definition and Aperture of Sieve Cloths

NF X 11-501 ET P 18-304		ASTM E 11-39 ET AFA		TYLER STANDARD SCREEN SCALE		BSA 410		DIN 1171	
Reference	Aperture	Reference	Aperture	Reference	Aperture	Reference	Aperture	Reference	Aperture
module	mm	no.	mm	mesh	mm	no.	mm	no.	mm
17	0.040	3.5	5.66	2½	7.925	300	0.053	0.060	0.060
18	0.050	4	4.76	3	6.680	240	0.066	0.075	0.075
19	0.063	5	4	3½	5.613	200	0.076	0.090	0.090
20	0.08	6	3.36	4	4.699	170	0.089	0.100	0.100
21	0.100	7	2.83	5	3.962	150	0.104	0.120	0.120
22	0.125	8	2.38	6	3.327	120	0.124	0.150	0.150
23	0.160	10	2	7	2.794	100	0.152	0.200	0.200
24	0.200	12	1.68	8	2.362	85	0.178	0.250	0.250
25	0.250	14	1.41	9	1.981	72	0.211	0.300	0.300
26	0.315	16	1.19	10	1.651	60	0.251	0.400	0.400
27	0.400	18	1	12	1.397	52	0.295	0.430	0.430
28	0.500	20	0.84	14	1.168	44	0.353	0.500	0.500
29	0.63	25	0.71	16	0.991	36	0.422	0.600	0.600
30	0.80	30	0.59	20	0.833	30	0.500	0.750	0.750
31	1	35	0.50	24	0.701	25	0.599	1	1
32	1.25	40	0.42	28	0.589	22	0.699	1.200	1.200
33	1.60	45	0.35	32	0.495	18	0.853	1.500	1.500
34	2	50	0.297	35	0.417	16	1.003	2	2
35	2.50	60	0.250	42	0.350	14	1.204	2.500	2.500
36	3.15	70	0.210	48	0.295	12	1.405	3	3
37	4	80	0.177	60	0.246	10	1.676	4	4
38	5	100	0.149	65	0.208	8	2.057	5	5
		120	0.125	80	0.175	7	2.411	6	6
		140	0.105	100	0.147	6	2.812		
		170	0.088	115	0.124	5	3.353		
		200	0.074	0150	0.104				
		230	0.062	170	0.089				
		270	0.053	200	0.074				
		325	0.044	250	0.061				
		400	0.037	270	0.053				
				325	0.043				
				400	0.038				

Starting at the top of a system of superposed sieves, the apertures are in a geometric progression, with the ratio r being less than 1.

According to the European norms NF, DIN and BSA, the ratio is:

$$r \# \frac{1}{\sqrt[10]{10}} = 0.8$$

For the American norms ASTM and TYLER:

$$r \# \frac{1}{\sqrt[4]{2}} = 0.84$$

Bibliography

[BAR 56] BARTH W., "Berechnung und Auslegung von Zykloabscheidern auf Grund nenerer Untersuchungen", *Brennstoff-Wärme – Kraft*, vol. 8, no. 1, p. 1, 1956.

[BIR 77] BIRSS R.R., GERBER R., PARKER M.R., "Analysis of matrix systems in high intensity magnetic separation", *Filtration et Séparation*, pp. 339–340, July/August 1977.

[BRA 65] BRADLEY, *The Hydrocyclone*, Pergamon Press, London, 1965.

[BRU 68] BRUN E.A., MARTINOT-LAGARDE A., MATHIEU J., *Mécanique des fluides*, Editions Dunod, Paris, 1968.

[BRU 70] BRUN E.A., MARTINEAU-LAGARDE A., MATHIEU J., *Mécanique des fluides*, Éditions Dunod, Paris, 1970.

[BRU 77] BRUNNER K.-H., MOLERUS O., "Partikclbewegung im Tellerseparator", *Verfahrenstechnik*, vol. 11, no. 9, p. 538, 1977.

[BRU 80] BRUNNER K.-H., MOLERUS O., "Investigation on separation efficiency of a plate separator", *Solids Separation Processes Symposium in Dublin*, April 1980, Publication of series no. 9 of the European Federation of Chemical Engineering, 1980.

[CAL 77a] CALVERT S., "How to choose a particulate scrubber", *Chemical Engineering*, vol. 84, no. 18, pp. 54–68, August 1977.

[CAL 77b] CALVERT S., "Get better performance from particulate scrubbers", *Chemical Engineering*, vol. 84, no. 23, pp. 133–148, October 1977.

[DAH 54] DAHLSTROM D.A., "Fundamentals and applications of the liquid cyclone", *Chemical Engineering Progress Symposium*, vol. 50, no. 15, p. 41, 1954.

[DAV 45] DAVIES C.N., "Definitive equations for the fluid resistance of spheres", *Proceedings of the Chemical Society*, vol. 57 part 4, no. 322, p. 18, 1945.

[DEU 22] DEUTSCH W., "Bewegung und Ladung der Elektrizitätsträger im Zylinderkondensator", *Annalen der Physik (Leipzig)*, vol. 68, pp. 335–344, 1922.

[DUR 99] DUROUDIER J.-P., *Pratique de la filtration*, Éditions Hermès, 1999.

[GOO 77] GOOCH J.P., MC DONALD J.R., "Mathematical modeling of fine particle collection by electrostatic precipitation", *AIChE Symposium Series*, vol. 73, p. 146, 1977.

[HAN 61] HANSBERG G., "Die Naßentstaubung in er Hüttenindustrie Entwicklungsmöglichkeiten", *Staub–Reinhaltung der Luft*, vol. 21, no. 9, pp. 418–425, 1961.

[KEL 52] KELSALL M.A., "A study of the motion of solid particles in a hydraulic cyclone", *Transactions of the Institution of Chemical Engineers*, vol. 30, p. 87, 1952.

[LAN 05] LANGEVIN P., "Une formule fondamentale de théorie cinétique", *Annales de chimie physique*, vol. 5, p. 245, 1905.

[LAW 77] LAWRENCE A.R., "Mineral processing methods: review and forecast", *Chemical Engineering*, pp. 102–110, 29 June 1977.

[LEU 72] LEUTERT G., BÖHLEN B., "Der räumliche Verlauf von elektrischer Feldstärke und Rauniladungsdichte im Platton-Elektrofilter", *Staub–Reinhaltung der Luft*, vol. 32, no. 7, p. 297, 1972.

[MCL 76] MCLEAN K.J., "Factors affecting the resistivity of a particulate layer in electrostatic precipitators", *Journal of the Air Pollution Control Association*, vol. 26, p. 866, 1976.

[MOT 77] MOTTOLA A.C., "Diffusivities streamline wet scrubber design", *Chemical Engineering*, pp. 77–80, 19 December 1977.

[MUS 67] MUSCHELKNAUTZ E., BRUNNER K., "Untersuchungen an Zyklonen", *Chemie Ingenieur Technik*, vol. 39, nos. 9–10, p. 531, 1967.

[MUS 72] MUSCHELKNAUTZ E., "Die Berechning von Zyklonabscheidern für Gase", *Chemie Ingenieur Technik*, vol. 44, p. 63, 1972.

[NOU 85] NOUGIER J.P., *Méthodes de calcul numérique*, Masson, Paris, 1985.

[NUK 39] NUKIYAMA S., TANASAWA Y., "Experimentation of atomization of liquids", *Nikon kikaï Gakkai Ronbunshu (Trans. Soc. Techn. Engrs. Japan)*, vol. 5, no. 18, pp. 68–75, 1939.

[PAU 32] PAUTHENIER M., MOREAU-HANOT M., "La charge des particules sphériques dans un champ ionisé", *Journal de physique et le Radium*, vol. 3, pp. 590–613, 1932.

[PER 73] PERRY R.H., CHILTON E.H., *Chemical Engineers' Handbook*, 5th edition, McGraw Hill, New York, 1973.

[POZ 57] POZIN M.E., MUKHLENOV I.P., TARAT E.YA., "Removal of dust from gases by the foam method", *Journal of Applied Chemistry of the USSR*, vol. 30, pp. 297–302, 1957.

[RIE 61] RIETEMA K., "Performance and design of hydrocyclones – I General considerations", *Chemical Engineering Science*, vol. 15, p. 298, 1961.

[ROS 32] ROSIN P., RAMMLER E., INTERMANN W., "Grundlagen und Grenzen der Zyklonentstaubung", *Zeitschrift des Vereines Deutscher Ingenieure*, vol. 76, p. 433, 1932.

[SAL 80] SALUDO R., MUNG R.J., *Solids Separation Processes Symposium held in Dublin*, Éditions Fédération Européenne de génie chimique, April 1980.

[SEM 77] SEMRAV K.T., "Practical process design of particulate scrubbers", *Chemical Engineering*, pp. 87–91, 26 December 1977.

[SMI 75] SMITH W.B., MCDONALD J.R., "Calculation of the charging rate of fine particles by unipolar ions", *Journal of the Air Pollution Control Association*, vol. 25, no. 2, p. 168, 1975.

[SPI 92] SPIEGEL M.R., *Formules et tables de mathématiques*, Serie Schaum, McGraw Hill, Paris, 1992.

[STA 65] STAIRMAND C.J., "Removal of grit, dust and fume from exhaust gases from chemical engineering processes", *The Chemical Engineer*, vol. 43, no. 10, pp. 310–326, December 1965.

[TAH 68] TAHERI M., CALVERT S., "Removal of small particles from air by foam in a sieve-plate column", *Journal of the Air Pollution Control Association*, vol. 18, pp. 240–245, 1968.

[TER 49] TER LINDEN A.J., "Investigations into cyclone dust collectors", *Proceedings of the Institution of Mechanical Engineers*, vol. 160, pp. 233–251, p. 1949.

[WHI 55] WHITE H.J., “Modern electrical precipitation”, *Industrial Engineering and Chemistry*, vol. 47, no. 5, pp. 932–939, 1955.

[WHI 74] WHITE H.J., “Resistivity problems in electrostatic precipitation”, *Journal of the Air Pollution Control Association*, vol. 24, no. 4, p. 314, 1974.

[ZEN 75] ZENZ F.A., “Size cyclone diplegs better”, *Hydrocarbon Processing*, vol. 54, no. 5, pp. 125–128, 1975.

Index

B, C, D

bubbles, 53, 55, 58, 60, 61, 65
by-pass effect, 35
centrifugal
  acceleration, 128, 137
  compressor, 61
  decantation, 125, 133, 135, 137
  filtration, 138
  force, 5, 103
  separation, 55, 56
Coriolis acceleration, 116, 126, 130, 134
Cunningham correction, 26, 115
drag
  coefficient, 48, 78
  force, 48, 56

E, F, G

electrofilter, 5–9, 21, 26, 34, 38, 42, 45, 46
electrostatic
  charge, 97
  dust remover, 46
  repulsion, 25
  separation, 36
entrainment, 46–50, 55, 58, 63, 68, 73, 90
flash drum, 47, 61, 64
  horizontal, 47, 52
  vertical, 47, 64, 65
friction, 74, 93, 105–108, 124, 125
gas cyclone, 111, 120, 125

H, I, M

hydrocyclone, 103, 111–115, 120, 122, 125
ion bombardment, 21, 24
Maxwell–Boltzmann, 22, 25

P, Q, R

pulverization column, 67, 89, 97
quenching, 70, 79, 99
Reynolds number, 56, 141
Rosin–Rammler, 62

S, V, W

scrubber
  air, 97
  perforated-plate, 97
  wet, 96–99
sparks, 8, 10, 38, 40, 97
Stokes' law, 26, 60, 128, 135
venturi, 67–71, 75, 81, 82, 86, 95, 96
wet aerosol, 124

Printed in the United States
By Bookmasters